NATIONAL DEFENSE RESEARCH INSTITUTE

A New Tool for Assessing Workforce Management Policies Over Time

Extending the Dynamic Retention Model

Beth J. Asch, Michael G. Mattock, James Hosek

Prepared for the Office of the Secretary of Defense

Approved for public release; distribution unlimited

The research described in this report was prepared for the Office of the Secretary of Defense (OSD). The research was conducted within the RAND National Defense Research Institute, a federally funded research and development center sponsored by OSD, the Joint Staff, the Unified Combatant Commands, the Navy, the Marine Corps, the defense agencies, and the defense Intelligence Community under Contract W74V8H-06-0002.

Library of Congress Cataloging-in-Publication Data

Asch, Beth J.
  A new tool for assessing workforce management policies over time : extending the dynamic retention model / Beth J. Asch, Michael G. Mattock, James Hosek.
    pages cm
  Includes bibliographical references.
  ISBN 978-0-8330-8137-7 (pbk. : alk. paper)
  1. United States—Armed Forces—Personnel management—Mathematical models.
  2. United States—Armed Forces—Personnel management—Evaluation. I. Mattock, Michael G., 1961- II. Hosek, James R. III. Title.

    UB323.A7695 2013
    355.6'10973—dc23
                                                                        2013028776

RAND OFFICES
SANTA MONICA, CA • WASHINGTON, DC
PITTSBURGH, PA • NEW ORLEANS, LA • JACKSON, MS • BOSTON, MA
DOHA, QA • CAMBRIDGE, UK • BRUSSELS, BE

www.rand.org

Preface

The dynamic retention model (DRM) is a state-of-the-art modeling capability that supports decisionmaking about workforce management policy. The DRM is quite general and can be applied to a wide variety of workforce contexts and a wide variety of compensation and personnel policies, though to date the focus has been on supporting military compensation decisions to sustain the all-volunteer force in the United States. Specifically, the DRM has allowed analysts to simulate the retention and cost effects of alternative military compensation policies.

While the DRM is an extremely powerful tool, a drawback in the use of the model to date is that it has only focused on the steady state. That is, implementations of the model to date show only the retention and cost effects of alternative policies when the entire workforce is under the new policy versus when the workforce is under existing policy. A steady-state model does not provide information on the transition to the new steady state or on the effects of the policy over time until the new steady state is reached. It also does not allow decisionmakers to assess the effects of alternative implementation strategies that would affect cost during the transition and how quickly the new steady state is reached. Filling this gap is a challenging task, but important because policymakers are generally concerned about not just the long-term effects of policies in the new steady state but also the short-term effects in the transition to the steady state.

The research presented in this report fills this gap. It extends the DRM to allow simulations of the effects of alternative policies both in the steady state and in the transition to the steady state. It also shows the effects of alternative implementation strategies and how different policies can affect how quickly the population and costs move toward the new steady state.

The description of the new model is highly technical. Nonetheless, the workforce policies analyzed in this report to highlight the versatility of the DRM will be well familiar to decisionmakers and analysts alike. The research should be of interest not only to the research community concerned with models to support workforce management but also to decisionmakers concerned about how to assess the short- and long-term effects of workforce management policies.

The research presented in this report was funded by the Gene Gritton Award for Innovation in Defense and National Security. The Gritton Award honors Gene Gritton, former Vice President of the RAND Corporation's National Security Research Division (NSRD). The purpose of the award is to stimulate new thinking about how to solve difficult problems and help RAND NSRD sponsors and clients get ahead of new challenges, focusing on research that has the potential to make substantial advances in an important defense or national security policy area. This research is the first to receive the Gritton Award.

This research was conducted within the Forces and Resources Policy Center of the RAND National Defense Research Institute, a federally funded research and development center sponsored by the Office of the Secretary of Defense, the Joint Staff, the Unified Combatant Commands, the Navy, the Marine Corps, the defense agencies, and the defense Intelligence Community.

For more information on the Forces and Resources Policy Center, see http://www.rand.org/nsrd/ndri/centers/frp.html or contact the director (contact information is provided on the web page).

Contents

Figures

Tables

Summary

The dynamic retention model is a state-of-the-art modeling capability that permits analysis of the effects on workforce size, experience mix, and cost of changes to compensation and personnel policies. Much of the empirical application of the model has been for the U.S. military. In the military context, the DRM is a behavioral model of each service member's decision to stay or leave the military where members are rational and forward-looking, differ in their preference for the military versus the civilian sector, and face uncertainty about future events that may cause them to value military service more or less than civilian life. To date, the DRM has been used to assess the effects of policy changes in the steady state. In the case of the military, where the typical military career is 30 years, it would take 30 years to reach the new steady state as a result of a policy change.

Policymakers are often concerned about the effects of a policy change in the transition to the steady state, i.e., during the 30-year period before the new steady state is reached, and how different implementation strategies can affect the 30-year time path. A common implementation strategy is to "grandfather" existing members so only new entrants are covered by any policy change. Grandfathering is often desirable because policymakers do not want to break the implicit contract with existing members and so wish to ensure that "promises are kept." This is exactly the implementation strategy and logic suggested by Secretary of Defense Leon Panetta in an August 19, 2011, interview with the *Army Times* when he was discussing possible future changes to the military retirement system:

> People who have come into the service and put their lives on the line, been deployed to the war zones, fought for this country, and who have been promised certain benefits as result of that—I'm not going to break faith with what's been promised to them (Tilghman, 2011).

The problem with this approach is that it can take a long time before the effects of a policy are realized. Policymakers must wait until existing members flow through and separate and new members get enough experience to be affected by the policy. One solution to this problem, as we describe below, is to grandfather existing members but also give them the choice to switch to the new system. By offering a choice, the shift to a new policy allows members under the existing policy to continue with it, or, if they prefer, to opt for the new policy. More people will be under the new system more quickly, if substantial numbers choose to switch, so it allows policymakers to move toward the steady state faster. Furthermore, faith has not been broken, and those who decide to change would do so only if they expect to be better off under the new policy.

Existing methodologies typically used to assess the transition phase and the effects of transition strategies are either severely limited or logically inconsistent. For example, personnel inventory projection models cannot be used to analyze the effects of allowing grandfathered members to switch to a new system because they do not include a model of decisionmaking that would logically allow members to change their behavior during the transition. Similarly, the so-called annualized-cost-of-leaving (ACOL) approach in which estimates of retention responsiveness to pay are used to simulate the retention effects of pay changes over some time period has been shown to be inconsistent with rational optimizing behavior and assumes away the possibility that individuals may change their mind when new information is revealed to them.

The DRM has neither of these disadvantages; it is logically consistent and can permit analysis of behavioral changes among incumbent members during the transition period. However, to date, few have actually used the DRM in this way, mostly because of the huge challenges in constructing a DRM that incorporates the transition to the steady state. The steady state version of the DRM is already extremely complex, in part because it keeps track of time in three different ways (time in the active component, time in the Reserve component, and total time elapsed). Extending it to include the transition period adds a fourth time dimension (time elapsed since a policy change occurred), substantially increasing the complexity of the model.

The research summarized in this document tackles this problem. We extend the mathematical model that defines the DRM to incorporate the fourth time clock. Specifically, we add a clock that accounts for the member's state when the policy occurs. We call this state the member's cohort, defined by the member's years of service (YOS) when the policy change occurs. We then use recent DRM parameter estimates for Army enlisted personnel to develop computer code that implements the extended model and permits us to simulate retention behavior for each cohort. Importantly, the extended model allows us to simulate both the retention behavior of each cohort over time and the retention behavior of all cohorts in the aggregate for each time period since the policy change occurred. Thus, the total force can be observed in each period as a force planner or programmer might want to see it. We can simulate retention behavior in the 30-year transition.

We demonstrate this capability with two examples. The first is a separation bonus paid to members who reach 11 YOS. The second is a reform to the military retirement system. We consider these examples because they represent the types of policies that are currently under consideration by policymakers, though the specific examples we consider are unlikely to ever be adopted (nor do we recommend that they be adopted). Their purpose is to illustrate our new capability.

We apply the extended capability to consider a number of implementation strategies. In the case of the separation bonus, we consider the effects of grandfathering existing members under the current policy and only requiring new members to be covered by the new policy. We also consider the effects of targeted grandfathering. In this case, only members with more than 5 YOS are grandfathered, and those with 5 or fewer years are automatically placed under the new policy. We contrast the results of these policies with the results when all members, both existing and new members, are automatically placed under the new system and there is no grandfathering. As expected, we find that grandfathering results in a slower time path than the immediate conversion case. That is, the effects on retention take much longer to be realized when members are fully grandfathered than when they are immediately placed under the new system. Targeted grandfathering results in a more intermediate pace of change.

In the case of retirement reform, we contrast the effects of an implementation strategy that fully grandfathers existing members under the current system with the effects of one that fully grandfathers them but also allows them to choose to switch to the new retirement system if they prefer to do so. New members are automatically moved to the new system. Incorporating this choice option required that we extend the DRM even further to allow each member the choice to switch to the new system if the value of their future career is greater by doing so, given their cohort and preference for military service. That is, we extended the DRM to incorporate not just the decision to stay or leave the military but also the decision to switch to a new compensation system if permitted to do so.

To exercise the model, we created an illustrative new retirement system. The new system was designed to maintain retention across years of service at the current levels and also to reduce cost. Figure S.1 shows our simulation results on the percentage of each cohort that chooses to switch to the new system at each year of service. We find that 100 percent of the first cohort (cohort = 1) participates in the new system because, by design, new entrants are automatically covered by the new system. However, we also find that almost 90 percent of cohorts 2-3 choose to participate. These represent existing members with between 2 and 3 YOS at the time the policy change occurred who opted to switch to the new system. However, older cohorts at the time of the policy change are less likely to switch. For example, about 50 percent of those in cohort 5 opt to switch, and nobody in cohort 9 opts to switch. The reason more senior members do not opt to switch is that they are close enough to the 20-year vesting point in the current system that staying in it is always more valuable to them than switching to the new system.

Figure S.1
Percentage of Personnel at Each YOS Who Participate in Retirement Reform by Cohort
(Defined by YOS at the Time of the Policy Change)

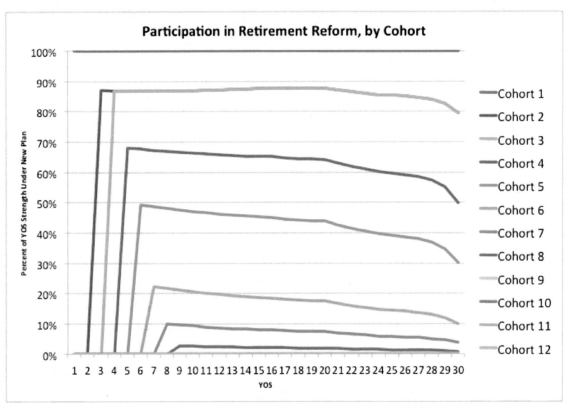

NOTE: Cohorts 9–12 are not seen in the figure because the participation rate is 0 percent.

Because we designed the retirement reform so that retention is unchanged—which maintains the size and experience mix of the force—one of the main differences between the two implementation strategies is in how quickly cost savings are realized. In the new steady state, we estimate that our retirement reform proposal would save about $1.8 billion annually for Army enlisted personnel. That is, the reform would allow the Army to sustain its current force size and experience mix at a cost of nearly $2 billion less per year, in the steady state. Figure S.2 shows the time pattern of reaching these cost savings under the fully grandfathered case versus when members have the option to switch to the new system. The figure shows the total cost of Army enlisted personnel under retirement reform as a percentage of the current baseline total costs when grandfathered members are permitted to switch to the new system and when they are not. As shown in Figure S.2, when members have the option to switch, the cost savings of the new policy are realized more quickly. Much of these cost savings occur because switching behavior enables the Department of Defense (DoD) to substantially reduce the accrual charge that is used to fund the military retirement system. The figure shows that the cost savings are greatest after 19 years have elapsed in the case of full grandfathering and is greatest after 14 years in the case where grandfathered members can switch. In the case of full grandfathering, DoD incurs the cost associated with paying transition pay to those covered by the new plan beginning in YOS 20. This additional cost stops the continuous decline in cost, and in fact, partially increases cost relative to the 19th year. In the case of grandfathering with switching, DoD incurs the costs associated with transition pay even earlier, after 14 years, when those who switched are eligible to receive this benefit.

Figure S.2
Total Personnel Costs Under Retirement Reform as a Percentage of Current Baseline Total Costs When Grandfathered Members Do and Do Not Have the Option to Switch, by Time Elapsed Since the Policy Change

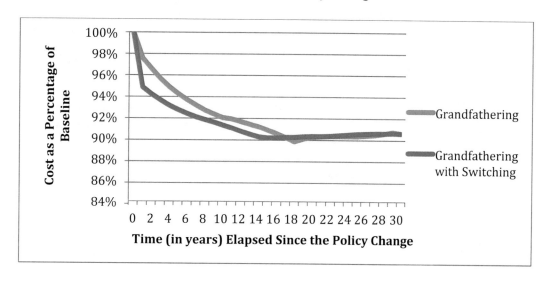

While the point of the analysis is not to argue for retirement reform or even a specific implementation strategy, our results do point to some clear policy results. We find that an implementation strategy that grandfathers existing members but gives them the choice to switch maintains the advantages of the full grandfathering policy, namely it allows policymakers to "keep faith" with current members and not break the implicit contract with respect to their compensation system. But unlike full grandfathering alone, allowing members to switch enables cost savings to be realized sooner than would otherwise be the case. Another advantage is that allowing choice is far more consistent with the current philosophy governing military manpower supply in the United States. That philosophy is one where military manpower is supplied by an all-volunteer force where people voluntarily choose to enter and stay in service. Thus, our analysis suggests that grandfathering with choice is an implementation strategy that should be given serious consideration if compensation reform is pursued in the future.

The capability to analyze the transition to a new steady state using the DRM approach represents a major step forward in the analytical tool kit available to researchers concerned with workforce management policy. While our analysis focuses on specific examples, the new capability has the potential to be applied to a wide variety of personnel and compensation policies as well as to workforces other than active duty personnel in the military. The capability is already being applied by related RAND projects.

Acknowledgments

We are extremely grateful to Gene Gritton and the Gene Gritton Award for Innovation in Defense and National Security committee for recognizing this research effort and to Jack Riley, Eric Peltz, John Winkler, and Jennifer Lamping Lewis at RAND for their input and support. We are honored to have been the first recipients of this award. Finally, we would like to acknowledge the three reviewers of the report, who provided extremely valuable comments, Kathleen Mullen and Bart Bennett at RAND and John Warner at Clemson University.

Abbreviations

AC	active component
ACOL	annualized cost of leaving
AHM	Asch, Hosek, and Mattock
CSRS	Civil Service Retirement System
DB	defined benefit
DC	defined contribution
DoD	Department of Defense
DRM	dynamic retention model
FERS	Federal Employees Retirement System
MHA	Mattock, Hosek, and Asch
QRMC	Quadrennial Review of Military Compensation
RC	reserve component
RMC	regular military compensation
s.d.	standard deviation
SRB	Selective Reenlistment Bonus
TSP	Thrift Savings Plan
YOS	year(s) of service

Section 1: Introduction

Recruiting and retention are the backbone of the all-volunteer force. Past research shows that a critical factor affecting the supply of personnel to the armed forces is the level and structure of military compensation relative to civilian alternatives. Military compensation is also a significant portion of the defense budget and affects the standard of living of military members.

Because of military compensation's importance as a recruiting and retention tool, a cost-driver, and a determinant of the financial well-being of military members, policymakers require information on how changes in the level and structure of military compensation will affect supply, cost, and the payout to military members. A challenge in meeting this requirement is that no data exist on policies that are yet untried. Researchers handle this challenge by estimating models of retention behavior using historical data and then use the estimated models to simulate how retention would change under alternative compensation policies. The dynamic retention model (DRM) is a state-of-the-art econometric model of officer and enlisted retention behavior in the active component (AC) and Reserve component (RC) and is the appropriate tool for assessing how retention and cost would change under alternative compensation schemes.[1]

The dynamic retention model was first developed by Glen Gotz and John McCall (1984) to study the retention of active duty Air Force officers. Since then the model has been applied more broadly to the entire enlisted force and officer force in the active as well as the Reserve components, to each service as well as to specific communities such as Air Force pilots and information technology personnel. The DRM has been extended to include the effect of compensation on the retention of higher-quality personnel and on incentives to exert effort and be productive.

However, in virtually all cases, past studies using the DRM approach consider only the effects of compensation policy in the steady state.[2] That is, they consider differences

---

[1] Past studies have primarily focused on compensation policy, but the DRM is also a tool for assessing the retention effects of personnel policies. Asch and Warner (2001) simulate the effects of changes in up-or-out policies, for example.

[2] The exception is Asch, Johnson, and Warner (1998), which uses the DRM approach to simulate both the steady-state effects and the effects in the transition to a new steady state of a specific military retirement reform proposal. Like the current study, it considers the case in which existing members can opt in to the new system. Unlike the current study, that study uses calibrated, rather than estimated, parameter estimates.

in retention behavior when all members are under the new compensation policy under consideration compared to when all members are under the current baseline policy.

These studies do not consider the transition to the steady state or policies that might influence that transition. This is a critical omission. In general, policymakers are concerned with both near-term and long-term objectives, e.g., increasing force size within a given number of years and ensuring that the resulting force reflects the desired long-term force structure. Furthermore, because these studies ignore the transition, existing DRM capabilities do not allow analysis of different implementation strategies. These strategies are temporary policies specifically designed to address how the military will transition from one steady state to the next. Thus, the analysis in these studies provides no information on how to get from "here to there" and the effects of implementation strategies to influence the path of from "here to there."

Addressing these concerns is complicated for several reasons. First, individual stay/leave decisions should be embedded in a multiperiod decisionmaking model, and in such a framework, retention is selective, which means that responsiveness to a policy change may differ by cohort depending on the number of years of service (YOS) at the time of the policy change. Second, the type and size of policy interventions can affect near- and long-term stay/leave behavior, policy cost, and the timing of outlays to implement the policy. A model should be capable of handling a wide variety of possible interventions, whether they are implemented singly or in a bundle, and showing their retention effects and cost by year. Policymakers should then have the information needed to weigh alternative policy approaches with respect to meeting near-term objectives, long-term objectives, and cost criteria. The present research is a step toward providing this capability through extending the DRM. The new capability enables users to analyze both the near- and long-term effects of policy changes in a unified framework.

Consider an example illustrating these concerns, namely, a hypothetical change that would make the military retirement system less generous but current compensation more generous, and suppose policymakers decided to "grandfather" existing members. That is, only new members would be covered by the new system, while existing members would be retained under the current system. Grandfathering is often desirable because policymakers do not want to break the implicit contract with existing members and so wish to ensure that "promises are kept." This is exactly the implementation strategy and the logic suggested by Secretary of Defense Leon Panetta in an August 19, 2011, interview with the *Army Times* when he was discussing possible future changes to the military retirement system:

> People who have come into the service and put their lives on the line,
> been deployed to the war zones, fought for this country, and who have

been promised certain benefits as result of that—I'm not going to break faith with what's been promised to them (Tilghman, 2011).

One challenge with grandfathering as an implementation strategy is that it can take years before the benefits of the permanent change are realized. The military retirement system vests personnel at 20 YOS with an immediate annuity, so grandfathering would mean that it would take 20 years before any member would retire under the new system. A 20-year horizon is well beyond the budget-planning horizon of most personnel planners.

Alternatively, suppose instead all existing members are immediately placed under the new system, regardless of YOS. Under this implementation strategy, members would be retiring under the new system immediately and so retention and cost effects of the new system would be felt immediately. It would also mean that existing members would be unprotected from the change to the new system and would have no time to anticipate and adapt to the changes before they occur. Some or many members might not prefer the new system to the baseline system, and this would be a clear case of breaking faith.[3]

Yet another alternative is to grandfather existing members but give them the choice to switch to the new system if they prefer to do so. This approach retains the advantage of grandfathering members. It also has the advantage that cost savings and retention changes are realized more rapidly if sufficient numbers of personnel switch over to the new system.

Thus, policymakers often face a tradeoff between the desire to protect existing members and extend the time horizon before any changes are actually realized against the desire to realize the benefits of change sooner but also expose existing members to unexpected and possibly unwanted change.

Extending the DRM to consider the transition to a new steady state is important because it allows policymakers to assess tradeoffs in terms of their near-term as well as long-term effects. That is, it provides policymakers with a capability to understand the workforce dynamics of different policies. It also provides policymakers with a tool to assess alternative implementation or transition strategies, such as grandfathering, grandfathering with choice, or immediate conversion, in terms of the tradeoff between the timing of when changes in cost and retention are realized and the effects on existing versus future members. Different implementation strategies can affect how quickly the new steady state is reached; the level and timing of personnel cost changes in transition to

---

[3] In other policy changes, costs might increase at first and then decrease in future years. For instance, suppose a separation incentive were offered to reduce the force size. Payment of the incentive would increase cost, but the resulting smaller force would cost less than the initial force.

the steady state; and whether, how, and when current members are exposed to the new policy. Other methods and models can be used to show the transition to the new steady state and the effects of alternative implementation strategies. However, as we discuss in Section 4 in detail, the DRM approach is superior to other approaches commonly used to study workforce dynamics.

However, extending the DRM to consider the transition to the steady state is a challenging task. It requires that we consider the time elapsed since the policy change occurred and that we simulate, year by year, how the force retention profile evolves over time. Because the DRM is a microsimulation model operating on the level of the individual service member, we must keep track of each member's status when the policy change occurs and then compute the impact on that status as time elapses. The DRM, even in its steady state version, is a highly complex model that allows members to be forward looking and heterogeneous in their tastes for active duty or Reserve military service, and it recognizes that they make decisions under uncertainty. This requires adding additional time dimensions that substantially increase the complexity of the model as well as the computational burden associated with running the model. Fortunately, advances in technology have made the computation requirements far less demanding in recent years, and the use of stochastic dynamic models of behavior has increased in the economics and management literatures.[4]

The research summarized in this report tackles this problem. It extends the steady-state DRM to incorporate the transition to the steady state. It also extends the DRM to consider the possibility of several different implementation strategies, including grandfathering of existing members, immediate conversion of existing members to the new policy, a gradual conversion plan, and a plan that grandfathers existing members but also gives them the choice to voluntarily switch to the new policy.

We demonstrate the new capability with two examples of hypothetical policy changes. The first is a change in current compensation and specifically a separation bonus paid to members who reach a specific point in their military career, namely 11 YOS. The second is a change in deferred compensation and specifically a new retirement system to replace the current one. The new system would be a defined contribution plan together with a less generous defined benefit plan, coupled with a lump-sum separation payment and additional continuation pay. The specifics of these options are discussed in Sections 2 and 3. We consider these hypothetical policies, not so much because we think

---

[4] See, for example, Hotz and Miller (1993); Rust (1994); Keane and Wolpin (1997); Aguirregabiria and Mira (2010); Bajari, Benkard, and Levin (2007); Van der Klaauw (2012); and Borkovsky, Doraszelski, and Kryukov (2012).

they are politically feasible changes or even desirable, but because they have clear retention or cost savings effects and highlight differences that can occur as a result of different transition policies. Furthermore, they represent the types of policies that are currently under consideration by policymakers. For example, the Army plans to drawdown its forces in the coming years, and separation bonuses are a potentially important tool that can affect the speed and cost of the transition to a smaller steady-state Army. Similarly, the Department of Defense (DoD) is undergoing a comprehensive review of military compensation, focusing on military retirement, and a range of polices to change the military retirement system is under consideration. Of particular interest are alternative systems that can achieve the same force size and shape but at a lower cost.

We also consider two examples, rather than just one, for practical reasons. The first example, a separation bonus, is a simple policy and easier to implement and illustrate in our computer modeling. The second example is much more complex, representing a fundamental change in the military compensation system, but also illustrates the type of policy change that the DRM is best adept at assessing.

Section 2 presents the extension of the DRM to the transition to the steady state. It also shows how we incorporate different transition strategies, such as grandfathering, and demonstrates the extended model with our first example, namely a separation pay at YOS 11. The section is technical and some readers may choose to skip it. Section 3 demonstrates the extended model with our second example, replacement of the current retirement plan with a new system. This demonstration shows how the new system can replicate the current retention profile and how different transition strategies can affect how quickly cost savings are realized. We present closing thoughts in Section 4. Finally, the Appendix provides additional model results regarding the retention effects of changing different model parameters.

Section 2: Extending the DRM to Incorporate the Transition Period

The DRM is an econometric model of officer and enlisted retention behavior in the AC and RC. It models service members as making retention decisions each year over their active and Reserve careers. The model assumes these members are rational and forward looking, taking into account both their own preference for military service and uncertainty about future events that may cause them to value military service more or less, relative to civilian life.

While the analyst cannot observe individual preferences for military service or the uncertainty assumed to affect decisions each period, we can assume the unknowns have known probability distributions. Using data on the stay/leave decisions of individual members, we can estimate the parameters of these probability distribution functions and the personal discount rate, and these estimated model parameters can then be used to simulate how retention behavior would change under alternative compensation policies, such as a change in pay or retirement benefits. For example, we can simulate the AC retention survival curve (the fraction of an entering cohort that stays to each year of service) under a new compensation policy and compare it to the simulated retention curve under the baseline current compensation policy. The survival curve in the steady state also gives the inventory of personnel at each year of service, given a steady force size, i.e., the retention profile for a given force.

In this section, we extend the DRM simulation capability to incorporate the transition to the new steady state.[1] Doing so requires that we introduce a new time dimension, namely time elapsed since the new policy is introduced so that we can simulate the retention profile by elapsed time until the new steady state is reached. Simulation of the retention curve by elapsed time involves extracting the relevant retention behavior from each cohort, where a cohort is defined as a group of members in a particular state when the policy occurs. In our analysis, cohort is defined by the member's year of service at the time of the policy change. For example, if a member has completed his or her 5th YOS when the policy is enacted, his or her cohort equals 5.

---

[1] The capability we describe in this section is a simulation capability using estimated parameters of the DRM. How we estimate the DRM parameters and the specific values of the estimated parameters are taken directly from Mattock, Hosek, and Asch (hereafter referred to as MHA) (2012) and Asch, Hosek, and Mattock (hereafter referred to as AHM) (2013). Readers interested in the data used, model estimation, and model fit are referred to these two reports.

The retention path of different cohorts may be affected differently by the new policy, especially if the implementation strategy of the new policy involves targeting the policy to some cohorts but not to others. For example, members who are currently in service at the time of the policy change may be required, or perhaps may be given the choice, to be covered by the new compensation policy. How they will be affected, and specifically their future retention path, will depend on where they are in their career when the policy is enacted. That is, the path of retention decisions over the career for a member who is at the beginning of his or her career may respond differently to a policy change than a member near the end or in mid-career. Alternatively, a new policy may fully "grandfather" existing members so the path is unchanged for those already in service. We consider these cases later in the report.

Our point here is that modeling the transition to the new steady state means that we must pay attention to the member's cohort, and specifically to his or her year of service, when the policy first occurs. We then use information on the cohort-specific retention decision paths to simulate how the retention profile evolves over time, as we describe below.

The section begins by incorporating cohort into the equations that define the DRM. In the process, we provide an overview of the steady-state DRM. The section then describes how we compute the retention profile by time elapsed since a policy is introduced. Next it discusses alternative implementation policies, such as "grandfathering" of current members, and how we incorporate these policies into the DRM. The section concludes with a brief summary.

Incorporating Cohort into the DRM

We follow past implementations of the DRM and assume that members start their military career on active duty. During each period of active service, the individual compares the value of staying in the AC with leaving and bases his or her decision on which alternative has the maximum value. The value of leaving is itself the maximum of choosing between the value of participating in the Reserves (including the option to leave and rejoin the Reserves at will, regulations permitting) and the value of leading a purely civilian life (with the option to participate in the Reserves at a later date, if eligible). If the member chooses to leave the AC, the member will select the alternative (Reserve or civilian) that yields the maximum value.

An individual choosing to stay in the AC can revisit the decision in each future period, so long as the individual continues to stay in the AC. An individual choosing to leave the AC will opt to be either a pure civilian or to participate in the Reserve, and the choice between civilian and Reserve can be revisited in each future period until retirement from the labor force, which is presumed to be age 60, or until the maximum combined active and Reserve career length is reached. All of these decisions will depend on his or her unique circumstances at a given point in time. Those circumstances include relative preference for AC or RC service to a purely civilian life and random events that may affect relative preferences over AC, civilian, and RC alternatives.

In the model, the value of staying depends upon the individual's preference for active military service (or "taste" for active service, which is assumed to be constant over time for a given individual), the compensation received for active service, and a period- and individual-specific environmental disturbance term (or "shock") that can either positively or negatively affect the value placed on active service in that period. For example, an unusually good assignment would increase one's relative preference for active service, while having an ailing family member who requires assistance with home care may decrease the value placed on active service. The value of staying also includes the value of the option to leave at a later date, that is, the individual knows that he or she can revisit the decision to stay or leave the next time it is possible to make a retention decision.

We make the simplifying assumption that once individuals have left active service, they do not reenter the AC. While there are instances where people do reenter the AC, the vast majority of those who leave do not reenter.

As mentioned, an individual who leaves the AC can choose to either be a civilian or combine civilian life with Reserve service. A person can join the RC immediately after leaving active service or can choose to join at a later date. Once a person enters the RC, he or she is free to choose to stay or to leave with the option of reentering at a later date, military regulations permitting.

At the beginning of each year of service, RC members compare the value of the civilian alternative—that is, leading a purely civilian life for that year—with the value of the Reserve alternative—that is, a first or additional year of Reserve service—and chooses the alternative that yields the maximum value.

The value of the civilian alternative includes the civilian wage, the AC or RC military retirement benefit the individual is entitled to receive (if any), an individual- and period-specific shock term that can either positively or negatively affect preference for the civilian alternative, and the future option to enter (or reenter) the RC, military regulations permitting.

The value of Reserve service includes the civilian wage; the Reserve compensation to which the individual is entitled, given his or her cumulative AC and RC service; an individual- and period-specific shock term; and the future option to either continue in the RC or return to a purely civilian life.

More technically, in each time period, the active service member compares the value of staying in the AC with the value of leaving and joining the RC or entering civilian life.

(Service members generally serve under a multiyear contract or have a multiyear obligation of military service, so our assumption allowing a stay/leave choice in each period may seem strong. We discuss this below after presenting the model's equations.) We use a nested logit approach to capture this decision, where the active service member is modeled as comparing active service to a civilian/Reserve nest.[2]

Active service has the value

$$V_a + \epsilon_a, \tag{1}$$

where V_a is the nonstochastic portion of the value of the active alternative, and ε_a is the environmental disturbance (shock) term specific to the active alternative, assumed to be extreme-value distributed. The value of active service varies with time, but we suppress the time subscripts in eq. (1).

The civilian/Reserve nest has the value:

$$max[V_r + \omega_r, V_c + \omega_c] + \upsilon_{rc} \tag{2}$$

where V_r is the nonstochastic portion of the value of the Reserve alternative, and V_c is the nonstochastic portion of the value of the civilian alternative; ω_r and ω_c are the shock terms specific to the Reserve and civilian alternatives, respectively, and υ_{rc} is the civilian/Reserve nest-specific shock.

The mathematical symbols for the DRM are summarized in Table 2.1.

---

[2] The Reserve/civilian choice is one nest and the active choice is the other nest. Under the assumption that the shocks have the same extreme value distribution, and, in particular, when they have the same variance, then the choice between the nests can be shown to have the usual logit form. Train (2003, Chapter 3) provides a proof that, when alternatives have identically distributed, independent extreme-value errors, the probability that a particular alternative is the maximum has the logit form. Ben-Akiva and Lerman (1985) show that the nested logit model can be written as a choice between alternatives, each of which is the maximum choice from its nest. See Asch et al. (2008) for application to the active, Reserve, and civilian choice in the military context.

Table 2.1
DRM Mathematical Symbols

Symbol	Meaning
V_a	Nonstochastic value of the AC alternative
V_r	Nonstochastic value of the RC alternative
V_c	Nonstochastic value of the civilian alternative
ε_a	Active alternative specific shock term, $\varepsilon_a \sim EV\left[o, \sqrt{\lambda^2 + \tau^2}\right]$
ω_r	Reserve alternative specific shock term, $\omega_r \sim EV[0, \lambda]$
ω_c	Civilian alternative specific shock term, $\omega_c \sim EV[0, \lambda]$
v_{rc}	Civilian/Reserve nest-specific shock term, $v_{rc} \sim EV[0, \tau]$
λ	Scale parameter of the distribution of ω_r and ω_c
τ	Scale parameter of the distribution of v_{rc}
γ_a	Taste for active service relative to civilian alternatives, $\{\gamma_a, \gamma_r\} \sim N[M, \Sigma]$
γ_r	Taste for Reserve service relative to civilian alternatives, $\{\gamma_a, \gamma_r\} \sim N[M, \Sigma]$
β	Personal discount factor, $\beta = 1/(1 + r)$ where r is the personal discount rate
W_a	Active compensation (RMC)
W_c	Reserve compensation
R	Military retirement benefit

Note: $EV[a,b]$ refers to the extreme value distribution with location parameter a and scale parameter b; the mean is given by $a + b\phi$ and the variance is given by $\pi^2 b^2/6$ where ϕ is Euler's gamma (~0.577). Similarly, $N[M,\Sigma]$ is the bivariate normal distribution with means given by the vector M and with covariance matrix Σ. The parameters M, Σ, λ, τ, and β are estimated using maximum likelihood methods. RMC refers to regular military compensation and includes basic pay, the basic allowances for housing and subsistence, and the tax advantage from receiving these allowances tax free.

The value of staying in the AC is the sum of the individual's taste for active service, γ_a; active military compensation, W_a; and the discounted value of the expected value of the maximum of the AC and civilian/RC nest alternatives in the following period. Note that to calculate wages, eligibility for retirement benefits, and so on, we need to keep track of time spent in the AC, time in the RC, and time overall. That is, the value of staying in the AC depends on these three time indices (time spent in the AC, time in the RC, and time overall). Furthermore, if a new policy is introduced the value of staying versus leaving may depend on member's state when the policy is enacted. In this version of the model, military pay, W_a, does not depend on cohort, but the model could be extended to permit this.

In our model, we denote the state when the policy is introduced by the variable "cohort" which is the member's year of service when the policy occurs. Thus, we have four time indices in the model. (In the steady state version of the DRM, there are only the first three indices). Note that each of the three indices are incremented appropriately to reflect the result of the choice in the following period as well as the member's cohort when the policy is introduced. For example, the value of staying on active duty is given by:

$$V_a(t_{active}, t_{reserve}, t_{total}, cohort) = \gamma_a + W_a(t_{active}) + \beta E[max[V_a(t_{active} + 1, t_{reserve}, t_{total} + 1, cohort) + \epsilon_a, max[V_r(t_{active}, t_{reserve} + 1, t_{total} + 1, cohort) + \omega_r, V_c(t_{active}, t_{reserve}, t_{total} + 1, cohort) + \omega_c] + v_{rc}]] \tag{3}$$

If the member leaves the AC, we assume he or she cannot return to the AC in any future period. The choice is then between serving in the RC (while holding a civilian job) or being a civilian and not serving at all. Those who leave the RC may later return. The value of the RC alternative, given the individual has left the AC, is the sum of the individual's taste for Reserve service, γ_r; Reserve military compensation, W_r; civilian compensation, W_c; and the discounted value of the expected value of the maximum of the civilian and Reserve alternatives in the following period:

$$V_r(t_{active}, t_{reserve}, t_{total}, cohort) = \gamma_r + W_c(t_{total}) + W_r(t_{active}, t_{reserve}) + \beta E[max[V_r(t_{active}, t_{reserve} + 1, t_{total} + 1, cohort) + \omega_r, V_c(t_{active}, t_{reserve}, t_{total} + 1, cohort) + \omega_c] + v_{rc}] \tag{4}$$

Finally, the value of the civilian alternative, given the member has left AC, is the sum of civilian compensation and the present value of any active or Reserve military retirement benefit for which the individual is eligible, W_c, and the discounted value of the expected value of the maximum of the civilian and Reserve alternatives in the following period:

$$V_c(t_{active}, t_{reserve}, t_{total}, cohort) = W_c(t_{total}) + \beta E[max[V_r(t_{active}, t_{reserve} + 1, t_{total} + 1, cohort) + \omega_r, V_c(t_{active}, t_{reserve}, t_{total} + 1, cohort) + \omega_c] + v_{rc}] \tag{5}$$

Note that to claim either an active or Reserve retirement benefit in our model, the member must have left service and be a civilian. Thus, military retirement benefits are incorporated into W_c and thus V_c in eq. (5) and thus in V_c in the $E[max[.]]$ expression in eqs. (3) and (4).

Derivation of Cohort-Specific Choice Probabilities and Retention Curves

As mentioned, as analysts, we do not observe individuals' tastes for active or Reserve service or the random shock terms. Instead, we assume they are each distributed according to known probability distribution functions with unknown parameters that we estimate using available data. Specifically, we assume individuals' tastes for active and Reserve service are bivariate normally distributed. Given these distributional assumptions, we can derive choice probabilities for each alternative at each decision year, the cumulative choice probabilities or survival probabilities for an entering cohort to each decision year, and write an appropriate likelihood equation to estimate the parameters of the model (the parameters of the probability distribution for the shock terms, the parameters for the population distribution of taste for active and Reserve service, and the discount factor). These derivations for the steady state DRM are documented in Asch et al. (2008). The data and estimation procedure used to generate the parameter estimates in the steady state model are documented in MHA (2012). Those studies also show how the retention survival curves simulated using the estimated parameters compare to observed retention patterns. In general, the model fit is quite good.

Extending the derivations of the choice probabilities and survival probabilities to include the transition to the steady state is straightforward. Since the value functions for the choice of AC, RC, or civilian status depend on cohort, so do the choice probabilities and the survival probabilities.

As shown in AHM (2008), given the nested logit assumption for the shock distributions, we can rewrite V_a, V_r, and V_c as:

$$V_a(t_{active}, t_{reserve}, t_{total}, cohort)$$

$$= \gamma_a + W_a(t_{active}) + \beta \kappa \left[\phi + ln \left[e^{\frac{V_a}{\kappa}} + \left[e^{\frac{V_r}{\lambda}} + e^{\frac{V_c}{\lambda}} \right]^{\frac{\lambda}{\kappa}} \right] \right]$$

$$V_r(t_{active}, t_{reserve}, t_{total}, cohort)$$

$$= \gamma_r + W_c(t_{total}) + W_r(t_{active}, t_{reserve}) + \beta \left[\phi \kappa + \lambda ln \left[e^{\frac{V_r}{\lambda}} + e^{\frac{V_c}{\lambda}} \right] \right]$$

$$V_c(t_{active}, t_{reserve}, t_{total}, cohort) = W_c(t_{total}) + \beta \left[\phi \kappa + \lambda ln \left[e^{\frac{V_r}{\lambda}} + e^{\frac{V_c}{\lambda}} \right] \right],$$

(6)

where $\kappa = \sqrt{\lambda^2 + \tau^2}$. Thus, we have explicit expressions for each value function, given (unobserved to the analyst) tastes for active and Reserve service, γ_a and γ_r. Later in this subsection, we describe how we handle unobserved tastes by "integrating out" this source

13

of heterogeneity. Furthermore, given the nested logit distributional assumption, we can write the probability that a member in a given cohort in decision year t_{active} YOS chooses to stay in the AC as:[3]

$$Pr(A)(t_{active}, t_{reserve}, t_{total}, cohort) = \frac{e^{\frac{V_a}{\kappa}}}{e^{\frac{V_a}{\kappa}} + \left[e^{\frac{V_r}{\lambda}} + e^{\frac{V_c}{\lambda}}\right]^{\frac{\lambda}{\kappa}}}.$$

(7)

To provide some intuition about this probability expression, it is straightforward to show that the probability of staying in the AC increases with V_a, the value of the AC alternative, and decreases with V_r and V_c, the values of the Reserve and civilian alternatives. That is, a member is more likely to choose the active alternative the higher the value of the active alternative and the lower the value of the Reserve and civilian alternatives.

The effect of an increase in the variance of the shock of the active alternative, κ in our model, depends on how V_a compares to the preferred choice in the Reserve-civilian nest, as we show in the Appendix. If V_a is smaller than the preferred choice, then an increase

---

[3] As mentioned, we assume service members can make a stay/leave choice in each period, whereas in reality they have multiyear commitments. This might affect our parameter estimates relative to estimating the model with full information about each member's current commitment. However, we think the effect of this is small for several reasons. In the early years of service, up to YOS 6, the estimated model includes switching costs to capture the difficulty (cost) of leaving prematurely. This helps to control for service commitments and should reduce their influence on estimates of the shock variance. In higher years of service when members have more military experience, their willingness to take on a multiyear service obligation reflects their ex ante assessment that the benefit of doing so is greater than the cost of not being able to leave at will. A mathematical version of this statement is in Mattock and Arkes (2007, pp. 12-13). So while members under contract are not at liberty in each period to leave, their willingness to enter into the contract implies that they do not expect the contract to be very constraining on their behavior—that is, they would expect to choose "stay" in each period if they were free to choose. It is those at the end of their contract who decide to stay or leave, and because contracts are of different length, in each period some will be making this decision. Our treatment of multiyear commitments might lead one to the conclusion that our estimate of the variance of the random shocks might be biased upward. For instance, if the shock variance had a negative effect on AC retention and if the contract effect caused retention to be higher than consistent with our assumption of free choice in each period, then the shock variance estimate might be biased upward. However, as discussed following equation (7), the effect of an increase in the shock variance (from an increase in κ) on $Pr(A)$ is ambiguous. An increase in κ decreases the probability only if the value of staying V_a is greater than the mode of the Reserve-civilian nest choice. This seems likely to be the case for senior personnel before 20 YOS because the expected value of military retirement benefits is high and increases V_a. But among junior personnel with a low taste for the military, V_a might be less than the mode, so an increase in κ would increase the probability of staying. In this case, one could argue that the variance estimate would be biased down.

in the variance of the shock increases the likelihood of choosing the active alternative, and if V_a is larger, an increase in the variance of the shock reduces the likelihood of choosing the active alternative. A further issue regarding the model is how the heterogeneity of taste for the military affects the responsiveness of overall retention to an increase in the value of the AC career, for instance from an across the board increase in military pay. An increase in the variance of taste can decrease this responsiveness, but depending on the point of evaluation, the reverse is also true. The Appendix discusses this point.

Returning to eq. (7), the notation means that for an AC member in cohort = 5 and decision year t_{active} = 6, Pr(A)(6,0,6,5) is the probability that a member with 5 YOS when the policy occurs decides to stay in the AC after completing 6 years of AC service. For $t_{active} \leq 5$, i.e., the decision years for this cohort prior to the introduction of the new policy, we assign the baseline steady-state probabilities. Thus, for cohort = 5, Pr(A)(1,0,1,5) is the baseline steady state probability that an AC member with 1 year of completed AC service stays in the AC. Note that for the cohort in the last year of the AC career when the policy occurs, cohort = 30, their probabilities for each decision year would be the baseline steady-state probabilities. That is, the baseline steady-state probabilities are those for cohort = 30. Similarly, the new steady-state probabilities are those for cohort = 0, i.e., new entrants who have not been covered by any policy except the new one.

If the member leaves, his or her choice in each future period is the decision to participate in the RC or leave the military altogether and be a civilian who does not serve. The probability a member who has left the AC decides to participate in the RC is:

$$Pr(R|RC)(t_{active}, t_{reserve}, t_{total}, cohort) = \frac{e^{\frac{V_r}{\lambda}}}{e^{\frac{V_r}{\lambda}} + e^{\frac{V_c}{\lambda}}}.$$

$$(8)$$

And similarly, the probability this member does not participate in the RC is:

$$Pr(C|RC)(t_{active}, t_{reserve}, t_{total}, cohort) = \frac{e^{\frac{V_c}{\lambda}}}{e^{\frac{V_r}{\lambda}} + e^{\frac{V_c}{\lambda}}}.$$

$$(9)$$

Note that the conditional probabilities of being Reserve and civilian in eqs. (8) and (9) sum to 1, given our assumption that the Reserve/civilian choice is nested.

We can also write the probability that a member who is still in the AC decides to leave and become a reservist. Letting "RC" represent the Reserve-civilian nest, this is given by:

$$Pr(A \text{ to } R)(t_{active}, t_{reserve}, t_{total}, cohort) = \big(1 - Pr(A)\big) * Pr(R|RC)$$

$$= \left(\frac{\left[e^{\frac{V_r}{\lambda}} + e^{\frac{V_c}{\lambda}} \right]^{\frac{\lambda}{\kappa}}}{e^{\frac{V_a}{\kappa}} + \left[e^{\frac{V_r}{\lambda}} + e^{\frac{V_c}{\lambda}} \right]^{\frac{\lambda}{\kappa}}} \right) \left(\frac{e^{\frac{V_r}{\lambda}}}{e^{\frac{V_r}{\lambda}} + e^{\frac{V_c}{\lambda}}} \right). \tag{10}$$

And similarly, we can write the probability the member on AC leaves and becomes a civilian:

$$Pr(A \text{ to } C)(t_{active}, t_{reserve}, t_{total}, cohort) = \big(1 - Pr(A)\big) * Pr(C|RC)$$

$$= \left(\frac{\left[e^{\frac{V_r}{\lambda}} + e^{\frac{V_c}{\lambda}} \right]^{\frac{\lambda}{\kappa}}}{e^{\frac{V_a}{\kappa}} + \left[e^{\frac{V_r}{\lambda}} + e^{\frac{V_c}{\lambda}} \right]^{\frac{\lambda}{\kappa}}} \right) \left(\frac{e^{\frac{V_c}{\lambda}}}{e^{\frac{V_r}{\lambda}} + e^{\frac{V_c}{\lambda}}} \right). \tag{11}$$

Because these probabilities depend on V_a, V_r, and V_c, and these values, in turn, depend on cohort, the probability a member at a given decision year (t_a, t_r, t_{total}) makes a particular choice depends on his or her cohort when the policy is enacted.

To understand how enactment of a policy affects the retention path of each cohort, defined by the member's year of service when the policy occurs, we must compute the survival curve or retention curve for each cohort. We start with the steady-state retention survival curve. In the steady state, the likelihood that an AC entrant survives and is still in the AC in decision year t_a is the probability the entrant stays in decision year 1, year 2, year 3, and so forth to year t_a. (Similarly, when an individual celebrates a birthday at age 5, the individual has completed 5 years of life as of the date of the birthday.) That is, it is the product of the probability of staying in the AC in each year to t_a. Similarly, for a given cohort, it is the product of the probabilities for that cohort, or:

$$cumulativePr(A)(t_a, t_r, t_t, cohort) = \prod_{s=1}^{t_a} Pr(A)(s, t_r, s, cohort). \tag{12}$$

For example, for a member in cohort = 5 when the policy occurs, the cumulative probability of staying in the AC to YOS 6, say, is:

$cumulativePr(A)(6,0,6,5) =$
$Pr(A)(1,0,1,5)Pr(A)(2,0,2,5)Pr(A)(3,0,3,5)Pr(A)(4,0,4,5)Pr(A)(5,0,5,5)Pr(A)(6,0,6,5).$

The cumulative probabilities to each decision year depend on the probabilities that, in turn, depend on the values V_a, V_c, and V_r. As shown above in equation (6), we have explicit expressions for these values, conditional on tastes for active and Reserve service. To compute these values we need to address the fact that taste is unobserved to the analyst. We handle this issue by taking advantage of our assumption that tastes have a bivariate normal probability distribution. Given this assumption, we can integrate the cumulative probabilities over the distribution of the tastes for active and Reserve services. In both the estimation and simulation, the integration is done numerically. As described in MHA (2012), for each individual, a Halton sequence of 23 pairs of active and Reserve seed tastes is drawn, and then, using trial values of the taste distribution parameters, the Halton draws are translated as though they were drawn from the distribution of tastes given the trial values of the parameters. The translation is done via a Cholesky decomposition (see Appendix B in AHM [2008]). For each resulting pair, the dynamic program is solved, giving values of the value functions at each decision point and hence values of the individual's career likelihood. The integration over tastes is accomplished by taking the average of the likelihoods over the 46 valuations.

The above expression (equation 12) for the cumulative probability illustrates the retention survival probability to each decision year for the cohort. Once we integrate out the heterogeneity in tastes for service, the sequence of cumulative probabilities to each decision year provides the retention survival curve for the cohort. That is, given the assumption of a maximum 30-year career in the AC, the AC retention curve for a given cohort is given by the vector:

$AC\ Retention\ survival\ curve(cohort)$
$= [cumulativePr(A)(1,0,1,cohort), cumulativePr(A)(2,0,2,cohort), ... , cumulativePr(A)(30,0,30,cohort)]$
$$(13)$$

This is the general expression for the cohort survival curve. However, we can simplify the computation of this curve for each cohort by taking advantage of the fact that the cumulative probability of retention to decision year t_a for cohort c is simply the baseline steady-state retention probability if $t_a \leq$ cohort. For example, as before, consider a member in YOS 5 when the policy is enacted (e.g., $c = 5$). When this member entered AC, his or her likelihood of being retained through decision years 1, 2, through YOS 5 is the steady-state baseline retention probability of being retained to each of these decision years. However, these likelihoods may differ from the steady-state likelihoods beginning in YOS 6 after the policy is enacted for this cohort. Such might be the case if the new policy immediately affects existing service members, including those in cohort 5. Thus, to compute the retention curve for cohort = 5 in our example, we would use the cumulative retention probabilities through year 5 under the baseline policy and use the cumulative retention probabilities under the new policy beginning in decision year 6 through year 30.

Let j denote the policy regime, with $j = 1$ denoting the current policy and $j = 2$ denoting the new policy. Thus, the retention curve for cohort = 5 is:

$$AC\ Retention\ survival\ curve(5) =$$
$$\begin{bmatrix} cumPr(A)^{j=1}(1,0,1,5), \dots, cumPr(A)^{j=1}(5,0,5,5), \\ cumPr(A)^{j=2}(6,0,6,5), \dots cumPr(A)^{j=2}(30,0,30,5) \end{bmatrix},$$

(14)

where

$$cumPr(A)^{j=2}(6,0,6,5) = cumPr(A)^{j=1}(5,0,5,5)\ Pr(A)^{j=2}(6,0,6,5).$$

More generally, for cohort = c, we have:

$$AC\ Retention\ survival\ curve(c) =$$
$$\begin{bmatrix} cumPr(A)^{j=1}(1,0,1,c), \dots, cumPr(A)^{j=1}(c,0,c,c), \\ cumPr(A)^{j=2}(c+1,0,c+1,c), \dots cumPr(A)^{j=2}(30,0,30,c) \end{bmatrix},$$

(15)

where

$$cumPr(A)^{j=2}(c+k,0,c+k,c) =$$
$$cumPr(A)^{j=1}(c,0,c,c)\ \prod_{s=c+1}^{c+k} Pr(A)^{j=2}(s,0,s,c),$$
$$\text{given } k \geq 1 \text{ and } c + k \leq 30.$$

Later in this chapter, we consider other implementation policies where current members are not automatically covered by the new policy ($j = 2$). The point we are illustrating with eqs. (14) and (15) is that for any cohort c we can use the baseline steady-state probabilities for decision years prior and through decision year equal to c.[4]

---

[4] While equations (13) and (15) provide general expressions of the AC cohort survival curve, we note that the expressions do not include the effects on cohort retention of changes in career opportunities produced by changes in the retention of other cohorts. For example, increases in separations among cohort 11 could change the promotion, pay, and therefore retention profile of earlier cohorts, such as cohort 8, 9, and 10. Incorporating these feedback effects requires a general equilibrium model of retention across cohorts and

Example

We demonstrate the cohort-specific retention curves with a policy example. The policy involves offering a separation incentive pay of $100,000 to enlisted personnel when they reach 11 YOS. That is, those with fewer or more than 11 years are not qualified for the incentive, but those with exactly 11 years can receive this benefit only if they choose to separate. We assume that all personnel who reach YOS 11 can expect to receive this incentive if they choose to leave. We can also consider the case where the policy is transitory so that only those who are YOS 11 *at the time of policy enactment* are eligible, but do not show those results here.[5] We illustrate the results of the example for Army enlisted personnel. The example is a simple policy and is meant to be illustrative. The next section considers more complex policies, focusing on retirement reform.

Figures 2.1 and 2.2 show the retention survival curves for selected cohorts and for each cohort, respectively, in this example. Figure 2.1 shows four different cohorts, namely those who were in YOS 1 when the policy was enacted ($c = 1$) as well those who were in YOS 11, 12, and 30 ($c = 11$, $c = 12$, and $c = 30$). The survival curves show how cumulative retention unfolds over the remainder of their career for each of these cohorts. These survival curves are compared to the baseline steady-state curve. The baseline survival curve is denoted as cohort = 0 or $c = 0$ in the figure. Figure 2.2 is three-dimensional and shows all cohorts; the horizontal axis is the member's year of service, the diagonal axis is the member's cohort when the policy is enacted (cohort spans from 1 to 30), and the vertical axis is the cumulative survival probability of staying in the AC until a given YOS.

Several findings are apparent. First, as expected, the retention of those with exactly 11 YOS when the policy occurs (i.e., cohort = 11 in Figures 2.1 and 2.2) have a dramatic drop in retention, relative to the baseline steady-state retention curve. Some individuals who might have stayed in the absence of the separation incentive are induced to leave the AC. Second, we find *increased* retention among those who had less than 11 YOS when the $100,000 separation pay policy was introduced (cohort = 1 in Figure 2.1 and cohort < 11 in Figure 2.2). This occurs because the DRM allows military personnel to be forward-looking and because we assume that any member who ever reaches YOS 11 will

---

would substantially increase the computational complexity of the model. It would also require us to develop a model of expectations formation and how changes in retention of other cohorts affect expectations. We believe extending the model in this way is feasible and desirable, but we leave such an effort for future research.

[5] The model assumes little or no advance notice of the policy. That is, policy responses are permitted only after the policy has been implemented. It is relatively straightforward to extend the model to allow for the effects of advance notice. However, we do not present that extension here.

receive the incentive. Individuals with fewer than 11 YOS anticipate that they can receive the separation incentive upon completing 11 YOS and so the value of staying in the AC increases, and thus retention increases. Third, those with more than 11 YOS when the policy is enacted (cohort = 12 and cohort = 30 in Figure 2.1 and cohort > 11 in Figure 2.2) have no change in their retention survival curve. Their curve is the same as the baseline steady-state curve. Finally, the retention curve for those in cohort 1 represents the steady-state retention curve under the new policy. Assuming the policy remains in place, all cohorts that enter after this cohort will have the same retention profile as this cohort. The change in the cumulative survival of an entrant through YOS 11 drops by 30 percent between the steady states, from 12.1 percent to 8.4 percent.

These results differ when policy eligibility is targeted to specific cohorts at the time of the policy change. For example, suppose only those at YOS 11 at the time of policy enactment are eligible to receive the incentive. That is, suppose those with fewer than 11 YOS are ineligible for the incentive when they reach YOS 11. In this case, there is no "anticipatory" increase in retention among those with fewer than 11 YOS. The retention survival curve for these cohorts is the same as it is for members with more than 11 YOS at the time the policy is enacted, namely the baseline steady-state retention curve. Thus, in this case, the only cohort affected by the policy is the YOS 11 cohort.

Figure 2.1
Retention Curves Under a $100,000 Separation Pay Policy, for Selected Cohorts at the Time of the Policy Change

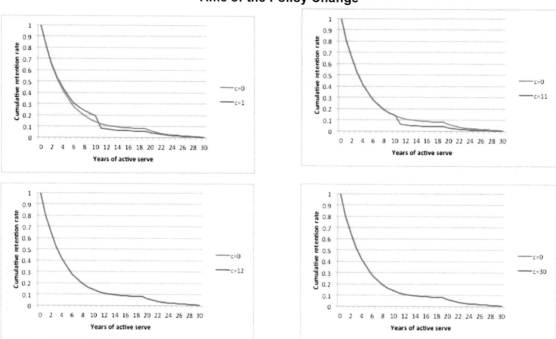

Figure 2.2
Retention Curves Under a $100,000 Separation Pay Policy, by Cohort at the Time of the Policy Change

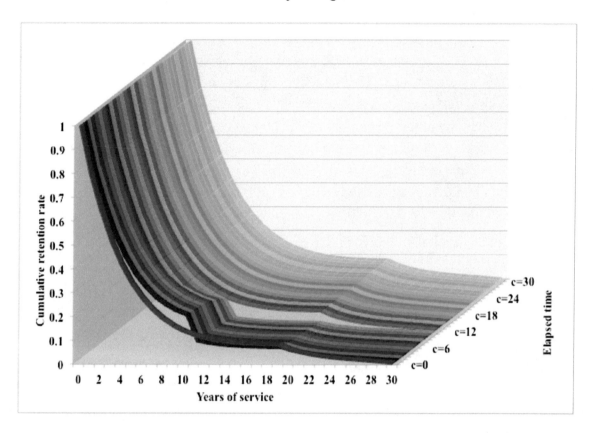

Also, given that the incentive to leave causes a drop in retention at YOS 11, it is likely that those choosing to stay had a high taste for the military, a high "positive" shock to remain in the military, or both. The average taste among those staying will be higher than it was in the absence of the incentive, and this will lead to higher retention from YOS 12 and beyond. In contrast, the higher average shock draw among those staying will not result in higher future retention; this is because shocks are not correlated between periods and affect only the current period.

This example shows that the retention effects of a policy in the transition to the new steady state vary by cohort. It also shows these effects can affect younger cohorts who anticipate a future benefit. We next show how these cohort-specific retention curves are used to generate retention profiles by time elapsed since the policy occurred. These retention profiles show the experience mix of the force (fraction of personnel at each year of service) for each year since the policy change.

21

Simulating the Evolution of Retention Effects over Time Due to a Compensation Policy Change

Simulating how the retention curve evolves over time in the transition to a new steady state when a compensation policy change occurs requires that we do a year-by-year simulation of the retention curve starting from the year in which the policy changed until the new steady state is reached. This simulation uses as input the cohort-specific retention survival curves described in the previous subsection.

Representing time properly requires careful attention. Consider the first year after the policy change is enacted. If s is time elapsed since the policy occurred, then $s = 0$ is the year when the policy occurred and $s = 1$ is the first year after the policy change. At time $s = 1$, those in YOS 2 were those who were in YOS 1 at $s = 0$. Similarly, those in YOS 3 at $s = 1$ were those in YOS 2 at $s = 0$. An alternative way to express this relationship is to say that at time $s = 1$, those in YOS 2 were in cohort $= 1$ at $s = 0$, and those in YOS 3 at $s = 1$ were in cohort $= 2$ at $s = 0$. Thus, the retention survival probability to YOS $= t_a$ at $s = 1$ is the survival probability of the t_a - 1 cohort, aged 1 year. More generally, the retention probability to YOS $= t_a$ at s is the survival probability for cohort $c = t_a$ - s, aged s years. Thus, elapsed time since the policy occurred is given by $s = t_a - c$.

We can illustrate how we select the relevant probabilities from each cohort retention survival curve as time elapses by considering Tables 2.2 and 2.3. The columns in Table 2.2 indicate YOS and are numbered from $t_a = 1$ to 30, where 30 is assumed to be the maximum length of an AC career. The rows indicate the cohort, defined by the year of service the member is in when the policy occurs, also spanning from 1 to 30. Each cell in the table represents the cumulative probability of reaching YOS $= t_a$ for cohort c, as defined by equation (12) above. For example, the element in the first row, fifth column is the cumulative probability that an entrant in cohort 1 stays in AC service through YOS $t_a = 5$. Similarly, the elements in the last column, are the cumulative probability that an entrant in the given cohort stays for a full 30-year career.

To illustrate how we compile the retention survival curves over time as s increases in Table 2.2, we show only the time clock s in Table 2.3 associated with the cumulative probabilities in Table 2.2. The baseline steady state is given by $s = 0$. We compile the baseline steady-state retention curve by selecting the diagonal elements in Table 2.2 and labeled $s = 0$ in bold font in Table 2.3. We compile the retention curve at $s = 1$ (1 year elapsed after the policy occurs) as the first off-diagonal elements in Table 2.2—the cells labeled $s = 1$ in Table 2.3. The cells that are in light font in Table 2.3—the lower

Table 2.2
How We Develop Retention Profiles by Time Elapsed: Cumulative Probabilities

	$t_a=1$	$t_a=2$	$t_a=3$	$t_a=4$	$t_a=5$	...	$t_a=29$	$t_a=30$
$c=1$	$\pi(1,0,1,1)$	$\pi(2,0,2,1)$	$\pi(3,0,3,1)$	$\pi(4,0,4,1)$	$\pi(5,0,5,1)$		$\pi(29,0,29,1)$	$\pi(30,0,30,1)$
$c=2$	$\pi(1,0,1,1)$	$\pi(2,0,2,2)$	$\pi(3,0,3,2)$	$\pi(4,0,4,2)$	$\pi(5,0,5,2)$		$\pi(29,0,29,2)$	$\pi(30,0,30,2)$
$c=3$	$\pi(1,0,1,1)$	$\pi(2,0,2,2)$	$\pi(3,0,3,3)$	$\pi(4,0,4,3)$	$\pi(5,0,5,3)$		$\pi(29,0,29,3)$	$\pi(30,0,30,3)$
$c=4$	$\pi(1,0,1,1)$	$\pi(2,0,2,2)$	$\pi(3,0,3,3)$	$\pi(4,0,4,4)$	$\pi(5,0,5,4)$		$\pi(29,0,29,4)$	$\pi(30,0,30,4)$
$c=5$	$\pi(1,0,1,1)$	$\pi(2,0,2,2)$	$\pi(3,0,3,3)$	$\pi(4,0,4,4)$	$\pi(5,0,5,5)$		$\pi(29,0,29,5)$	$\pi(30,0,30,5)$
...								
$c=29$	$\pi(1,0,1,1)$	$\pi(2,0,2,2)$	$\pi(3,0,3,3)$	$\pi(4,0,4,4)$	$\pi(5,0,5,5)$		$\pi(29,0,29,29)$	$\pi(30,0,30,29)$
$c=30$	$\pi(1,0,1,1)$	$\pi(2,0,2,2)$	$\pi(3,0,3,3)$	$\pi(4,0,4,4)$	$\pi(5,0,5,5)$		$\pi(29,0,29,29)$	$\pi(30,0,30,30)$

Note: The cells in the table are cumulative probabilities, where each probability is given by equation (12), where $\pi(t_a,t_r,t_t,c) = \text{cumulativePr(A)}(t_a,t_r,t_t,c)$ in equation (12). The columns are year of active service t_a and the rows indicate cohort c.

Table 2.3
How We Develop Retention Profiles by Time Elapsed: Value of the *s*
(Time Elapsed)

	$t_a=1$	$t_a=2$	$t_a=3$	$t_a=4$	$t_a=5$	$t_a=6$	$t_a=7$	...	$t_a=29$	$t_a=30$
$c=1$	$s=0$	$s=1$	$s=2$	$s=3$	$s=4$	$s=5$	$s=6$		$s=28$	$s=29$
$c=2$	$s=0$	$s=0$	$s=1$	$s=2$	$s=3$	$s=4$	$s=5$		$s=27$	$s=28$
$c=3$	$s=0$	$s=0$	$s=0$	$s=1$	$s=2$	$s=3$	$s=4$		$s=26$	$s=27$
$c=4$	$s=0$	$s=0$	$s=0$	$s=0$	$s=1$	$s=2$	$s=3$		$s=25$	$s=26$
$c=5$	$s=0$	$s=0$	$s=0$	$s=0$	$s=0$	$s=1$	$s=2$		$s=24$	$s=25$
$c=6$	$s=0$	$s=0$	$s=0$	$s=0$	$s=0$	$s=0$	$s=1$		$s=23$	$s=24$
$c=7$	$s=0$	$s=0$	$s=0$	$s=0$	$s=0$	$s=0$	$s=0$		$s=22$	$s=23$
...										...
$c=29$	$s=0$	$s=0$	$s=0$	$s=0$	$s=0$	$s=0$	$s=0$	$s=0$	$s=0$	$s=1$
$c=30$	$s=0$	$s=0$	$s=0$	$s=0$	$s=0$	$s=0$	$s=0$	$s=0$	$s=0$	$s=0$

Note: The table cells only show s, the time-elapsed clock for the cumulative probability in Table 2.2, where s is defined as $t_a - c$. The columns are year of active service t_a, and the rows indicate cohort c.

diagonal section of the table—are probabilities that are for career segments that occur before the policy was enacted. For example, YOS 1 through 3 for cohort = 4 (e.g., those with 4 YOS when the policy was enacted) occur prior to the policy. Therefore, we assign the baseline steady-state retention survival probabilities to these cells. These are also denoted as $s = 0$ in Table 2.3 (in light font).

Example

We illustrate how we compile the retention profile by elapsed time using the above example, namely a $100,000 separation incentive for those who ever reach YOS 11.

Figure 2.3 shows the retention profiles for selected elapsed times ($s = 1$, $s = 6$, $s = 12$, and $s = 30$ compared with $s = 0$) and Figure 2.4 shows the retention profiles for each elapsed time (from $s = 0$ to $s = 30$). Figure 2.4 is three-dimensional with the horizontal axis indicating year of service, the diagonal axis indicating elapsed time s since the policy occurred, and the vertical indicating the cumulative survival probability that an entrant reaches each year of service. The difference between Figures 2.1 and 2.2 versus Figures 2.3 and 2.4 is that the former figures show the retention profile by cohort while the latter ones show the retention profile by elapsed time, where the latter draws information on retention across cohorts, as in Tables 2.2 and 2.3. Figures 2.3 and 2.4 are the traditional retention profile over time, but unlike a mechanistic Markov process used to generate the retention probabilities, the probabilities are based on dynamic behavioral responses of individual members.[6]

In the first year after the policy change ($s = 1$), we see a large drop in retention at YOS 11 as more people leave to claim the $100,000 benefit. We also see an increase in retention among those with less than 11 YOS as those with fewer than 11 YOS anticipate getting the separation pay, leading to an increase in their retention. In the second year after the policy change ($s = 2$), we see a drop in retention among those with YOS 12. This occurs because fewer personnel remain after YOS 11, due to the separation incentive; the remaining group is smaller so the number of personnel in service at YOS 12 is correspondingly smaller than it was the year before. Similarly, in the second year after the policy, we see an increase in retention at YOS 11 (relative to the first year, $s = 1$), because a larger cohort of personnel with 10 YOS chose to stay in service an additional year and as this cohort ages 1 year, the group size is now larger relative to the previous year. We also see a further increase in retention among those with fewer than

---

[6] In Section 4, we contrast our approach with using a Markov process to model retention in the transition to the steady state.

11 YOS as the larger cohorts induced by the policy age 1 year. Similar effects are seen in subsequent years after the policy (such as $s = 6$ in Figure 2.3). The smaller YOS 11 cohort ages each year and as this smaller group flows through its career, retention is lower than it had been before the new policy. The larger pre-YOS 11 cohorts also age each year, and as this larger group flows through, it partially offsets the drop in retention due to the flow through of the smaller YOS 11 cohort. After 30 years, all cohorts have flowed through, and the new steady state is reached.

Figure 2.3
Retention Curves Under a $100,000 Separation Pay Policy, for Selected Elapsed Times Since the Policy Change

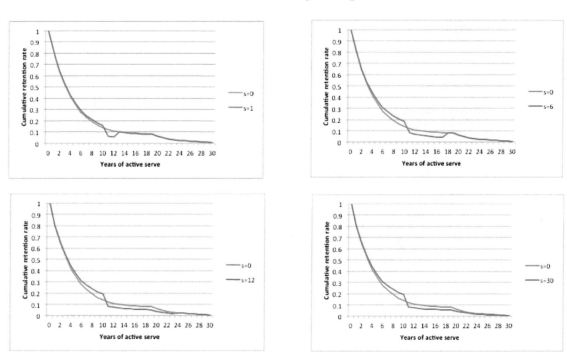

The results in Figures 2.3 and 2.4 illustrate several aspects of the model. First, the year-by-year retention profile in transition to the new steady state reflects the underlying effects on each cohort of the policy change. Second, the year-by-year change reflects the subsequent flow over the career of these cohorts. Thus, if a policy change reduces the retention of some cohorts and therefore the size of those cohorts, then the year-by-year changes will show the flow-through of those smaller cohorts over time until they complete their military career. Third, the initial effect of a policy might be larger than the steady-state effect. In the example shown in Figures 2.3 and 2.4, the drop in retention at YOS 11 after the first year ($s = 1$) was larger than the steady-state drop at that point ($s = $

25

30) because the initial drop was partially offset in later years by the larger, younger, cohorts as they aged and flowed through this point in their careers. That is, the trough that appears initially is dissipated over time as the larger, younger cohorts age and flow through their careers.

Figure 2.4
Retention Curves Under a $100,000 Separation Pay Policy, by Time Elapsed Since the Policy Change

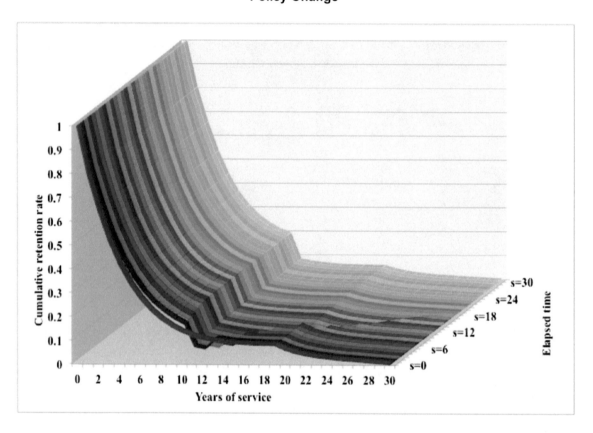

Incorporating Alternative Implementation Policies into the DRM

Compensation policy decisions involve not only decisions about how to change compensation but also how to implement the policy. The implementation approach or transition policy can affect who is affected by the new policy, how quickly the new steady state is achieved, and the trajectory of personnel costs in transition to the new steady state.

For example, a policy that grandfathers current members would allow existing members at the time of the policy change to remain under the current policy and require only that new entrants be transitioned to the new policy. This policy "protects" current members from the policy change and keeps faith with those members who chose to serve

26

in the military under the expectation that they would be covered by only the current compensation system. On the other hand, a policy that grandfathers current members means that the new steady state when all members are under the new policy will not be achieved for at least 30 years, when all current members have flowed through the system. Thus, a policy that gives maximum protection to current members results in a long time horizon for the effects of the policy change to be felt, and therefore a long horizon for the full cost savings to be realized.

The other extreme transition policy is to require all current members to immediately transition to the new policy. Under this transition policy approach, the new steady state and the full cost savings of the new steady state are realized very quickly, but this approach breaks faith with current members and provides no protection for those who do not wish to make a change. There are also intermediate approaches. One such approach would require only junior personnel to switch immediately to the new system, while grandfathering more senior personnel.

Another intermediate approach would be to grandfather all current members but allow these members to voluntarily switch to the new system during a specified enrollment window, say 1 or 2 years. Giving members a choice increases their opportunity set, and they can take the system that is better for them individually. For instance, a change in compensation that increased pay in more senior years of service would be more attractive to members with a high taste for military service as they would be more likely to have a long career. The approach of grandfathering all members but allowing switching has a number of advantages. First, it would clearly make existing members better off to the extent that those who would prefer the new system can switch and those who prefer the existing system can remain with it. Second, as a voluntary system, it is consistent with the choice-based system that is the foundation of the all-volunteer force. Third, it offers a compromise between the two extreme implementation approaches. It allows the new steady state and associated cost savings to be realized sooner than 30 years while still affording the protection to current members that would occur under a grandfathering policy with no voluntary switching.

The remainder of this subsection incorporates each of these transition policies into the DRM equations.

Two Extreme Transition Policies: Immediate Conversion of Current Members Versus Full Grandfathering

Consider the AC retention survivor curve for a given cohort shown in eq. (15). As discussed earlier, this expression shows the vector of cumulative probabilities from entry to each year of service for a given cohort where it is assumed that the cohort is immediately converted to the new policy. To show how cumulative probabilities differ depending on the transition policy, we can generalize eq. (15) as:

$$AC\ Retention\ survival\ curve(c) =$$
$$\begin{bmatrix} cumPr(A)^1(1,0,1,c), \dots, cumPr(A)^1(c,0,c,c), \\ cumPr(A)^j(c+1,0,c+1,c), \dots cumPr(A)^j(30,0,30,c) \end{bmatrix}.$$

(16)

In the case of immediate conversion of current members to the new policy, $j = 2$ in equation (16) for all cohorts c, and equation (16) is the same as equation (15).

In the case of full grandfathering of existing members, the new policy is relevant only for the entry cohort (and subsequent future entry cohorts), or $j = 2$ only for $c = 0$ in eq. (16). Current members are automatically covered by the current policy, so $j = 1$ for $c > 0$ in eq. (16).

The top part of Table 2.4 summarizes the rules for equation (16) under the two approaches of immediate conversion and of full grandfathering of current members.

Table 2.4
Alternative Transition Policy Rules for the DRM (Equation 16)

Transition Policy	Cohort c	Policy index j (j=1 is current policy, j=2 is new policy)
Immediate conversion of existing members		
	c≥0	j=2
Grandfathering of all existing members		
	c=0	j=2
	c>0	j=1
Targeted grandfathering of members in cohorts c>x		
	0≤c≤x	j=2
	c>x	j=1
Grandfathering of all members with choice to switch to new policy during switching window t_a=c		
	c=0	j=2
	c>0	Choose j=1 if $V^1_a > V^2_a$
		Choose j=2 if $V^2_a > V^1_a$

We can also illustrate these alternative transition policies in terms of elapsed time since the policy occurs. Figure 2.5 shows the transition to the new steady state for selected elapsed times ($s = 1$, $s = 6$, $s = 12$, and $s = 30$, compared to $s = 0$), while Figure 2.6 shows the transition for each elapsed time, assuming full grandfathering for the policy example we have considered so far, namely a $100,000 separation payment for those ever reaching YOS 11. In the case of full grandfathering, only entry cohorts will be affected by this policy, so the first time the separation incentive will ever be paid will be

28

11 years after the policy was enacted when the first entry cohort eventually reaches YOS 11. As the figures show, the effect of the policy occurs gradually over time as entry cohorts that are covered by the new policy age, new cohorts enter, and more senior uncovered cohorts get older and eventually leave the military. The entire force is covered by the new policy only after 30 YOS.

Earlier in Figures 2.3 and 2.4, we showed the transition approach that immediately converts existing members to the new policy. Comparing those figures with Figures 2.5 and 2.6, we see that immediately transitioning current members to the new policy (Figures 2.3 and 2.4) results in immediate changes throughout the force in the first year of the policy. That is, in Figures 2.3 and 2.4, we observe both the anticipatory positive retention effects in the pre-YOS 11 years and the negative post-11 retention effects in the first few years after the policy. In contrast, in Figures 2.5 and 2.6, we observe only a small uptick in retention in the first few years after the policy, as members anticipate the incentive at YOS 11.

Figure 2.5
Retention Curves Under a $100,000 Separation Pay Policy with Full Grandfathering of Existing Members, for Selected Elapsed Times Since the Policy Change

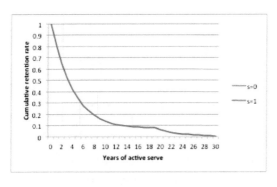
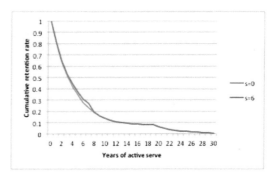
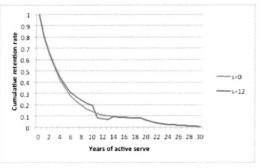
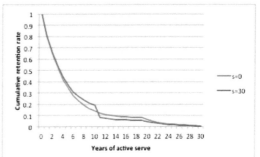

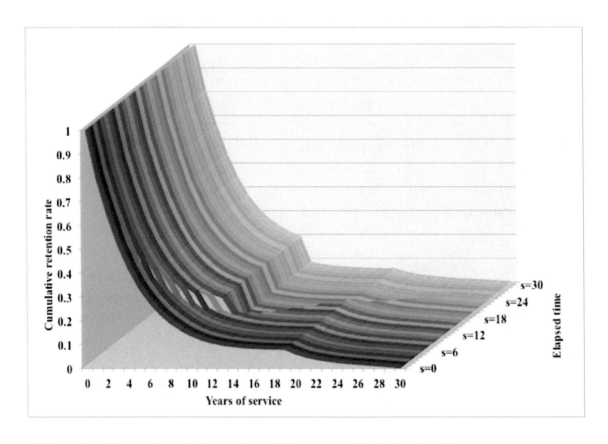

A Targeted Transition Policy: Grandfathering of A Subset of Members

An alternative implementation strategy is to grandfather only members who are more invested in the current system, namely senior members, and to require only junior members to be covered by the new policy. This approach was used to implement the transition of federal civil service employees to their new retirement plan in 1987, called the Federal Employees Retirement System (FERS). The predecessor of FERS was the Civil Service Retirement System (CSRS). The FERS implementation plan required that employees with 5 or fewer YOS be covered by FERS, while those with more than 5 years were grandfathered under CSRS. As part of the implementation strategy some CSRS-covered employees were allowed to voluntarily switch to FERS.[7] We consider the case of

---

[7] A discussion of the FERS transition policy and the financial incentives to switch from FERS to CSRS is provided in Asch and Warner (1999). Some CSRS-covered members can still switch to FERS. For

voluntary switching in the next subsection. Here, we consider the case of a targeted transition policy without switching.

More specifically, suppose a compensation policy is implemented in a targeted way so that only cohorts with x or fewer YOS are covered by the new policy while those with more than x YOS are grandfathered under the baseline current compensation policy. For cohorts with x or fewer YOS, the expression for the AC retention curve in equation (16) becomes equation (15) above. For cohorts with more the x YOS at the time of the policy, expression (16) becomes:

$$AC\ Retention\ survival\ curve(c) = \\ \begin{bmatrix} cumPr(A)^1(1,0,1,c), ..., cumPr(A)^1(c,0,c,c), \\ ccumPr(A)^1(c+1,0,c+1,c), ... cumPr(A)^1(30,0,30,c) \end{bmatrix}.$$

(17)

The rules for this transition policy are shown in Table 2.4.

Figures 2.7 and 2.8 show the transition path under this implementation policy for our example of a $100,000 separation incentive to those ever reaching YOS 11, assuming $x = 5$. The negative retention effects of the policy at YOS 11 are only revealed as time elapses and cohorts with 5 or fewer YOS get more senior and reach YOS 11. Thus, the first time we observe a drop in retention at YOS 11 is after 6 years have elapsed when those with 5 YOS at the time of the policy reach YOS 11. This is in contrast to the full grandfather case in Figures 2.5 and 2.6, where there is no drop in retention at YOS 11 until 10 years have elapsed and those who were entering YOS 1 at the time of the policy reach YOS 11.

Notice that because we assume people anticipate the policy action at YOS 11 (because we assume the policy is a permanent feature of the compensation system), retention increases among those with fewer than YOS 11. In the case of full grandfathering, this occurs very gradually, but in the case of grandfathering only, those with more than 5 YOS, the effect is still gradual, but less so. For example, after 1 year ($t + 1$ or $s = 1$ in the figures), retention increases for those with fewer than 6 YOS in Figures 2.7 and 2.8, the case of partial grandfathering, but increases only for those with fewer than 2 YOS in Figures 2.5 and 2.6, the case of full grandfathering.

---

example, CSRS-covered members who have a break in service and return to the federal civil service are permitted to switch to FERS.

An alternative assumption is that the policy is not permanent and not anticipated by personnel. Such might be the case for a policy intended to have a temporary effect, such as separation bonuses or retention bonuses intended to draw down a force or increase it. The DRM can accommodate a policy that is temporary (i.e., only available for a limited time), that is targeted to a specific group, or that is unanticipated. For instance, personnel at YOS 8, 9, and 10 might be offered a separation bonus in the current year and the next year. Those at YOS 8 could take the bonus and leave, or stay and take the bonus at YOS 9, or stay at YOS 9 and have no further separation incentive, and similarly for those at YOS 9 and 10 when the new policy goes into effect. Current RAND analysis is pursuing this approach as an extension of the analysis presented here.

Figure 2.7
Retention Curves Under a $100,000 Separation Pay Policy with Grandfathering Targeted to Cohorts with More than 5 YOS, for Selected Elapsed Times Since the Policy Change

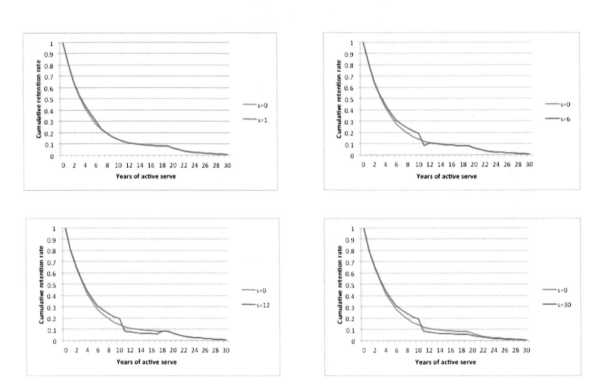

Figure 2.8
Retention Curves Under a $100,000 Separation Pay Policy with Grandfathering Targeted to Cohorts with More than 5 YOS, by Time Elapsed Since the Policy Change

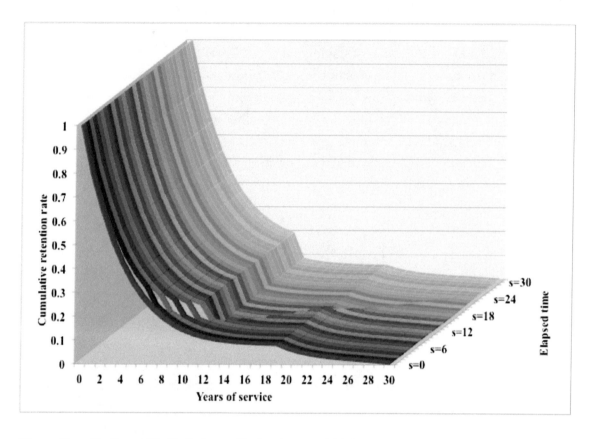

Transition Policy with Full Grandfathering and Voluntary Switching

Another intermediate implementation approach that also allows changes to occur sooner than the full grandfathering case is to grandfather all current members under the baseline current policy but also allow members to switch to the new policy during a specified enrollment period, such as one year. From the standpoint of the DRM, this approach means that we must model the member's choice and then keep track of that choice in computing the state probabilities, cumulative probabilities, and retention survival curves.

Suppose a new policy is introduced that grandfathers existing members, but these members have 1 year to choose to switch to the new policy. The choice of a 1-year window is arbitrary, and the model can be extended to consider switch windows of *n* years. We assume that once a member switches, he or she cannot switch back to the current system. This is consistent with how the FERS/CSRS choice was implemented. Those covered under CSRS who chose to switch to FERS are not permitted to switch back. An AC member with t_{active} YOS at the time of the policy is in the cohort equal to t_{active}, and our assumption of a 1-year window means that this member can make the choice only when t_{active} = cohort = c.

This AC member with t_{active} YOS will choose to switch to the new policy if the value of staying in the AC under the new policy is greater than the value under the existing policy. Put differently, the value of staying in the AC is the maximum of the value of staying under the new versus the existing policy. Following the notation we have used above, we denote the value of staying for a member with c years of AC service as $V_a(c,0,c,c)$, i.e., the value of staying active for a member with t_{active} = c years, no RC years, c total years, and for cohort = c. For members who choose to stay under the existing baseline policy, the value of staying for this cohort is $V^{j=1}_a(c,0,c,c)$, while for those who switch, the value of staying is $V^{j=2}_a(c,0,c,c)$. Thus, the value of staying for an AC member is:

$$V_a(c,0,c,c) = max(V_a^1, V_a^2).$$

Given the members' choices, their future values and probabilities will depend on the choice they made during the switching window. For those who switch to the new policy, the retention curve of eq. (16) becomes that of eq. (15). For those who choose to stay with the existing policy, the retention curve of eq. (16) becomes that of eq. (17).

We illustrate the case of grandfathering with choice in Section 4, where we consider a military retirement policy application. As we discuss in that section, we find that under this transition policy, the effects of the policy occur earlier than in the case of full grandfathering in the absence of choice but later than in the case of immediate conversion to the new policy.

Summing Up

To summarize, we extended the DRM capability to incorporate simulations of the transition to the steady state. We use our DRM model estimates (obtained from previous studies) to simulate the new retention profile for each cohort, defined by their YOS when the policy occurred, and then combine these profiles to derive the retention profile as time elapses. We considered alternative implementation or transition policies defined by whether and how current members are covered by the new policy.

We illustrated the capability with a simple policy—a $100,000 separation payment targeted to those who ever reach YOS 11. This policy is not realistic in the sense that it is highly unlikely policymakers would ever adopt such a policy for enlisted members. But it can be extended to more complex and more realistic policies. For example, we could consider a policy that targets multiple YOS. We can also consider policies that are not permanent but offered for only a limited period of time, say 1 or 2 years. That is, a service wanting to down size might offer separation incentives only until sufficient numbers of personnel leave and then stop the offer. A service wanting to increase experience in certain years of service and particular career fields might offer retention bonuses until sufficient numbers of personnel stay and then stop the offer. We can also consider policies where the incentive is not a lump sum, but based on years of service or monthly basic pay, for example. These are relatively straightforward extensions. Other research under way at RAND is conducting such extensions of this capability.

In the next section, we consider a more complex policy, and one that has been the subject of much debate, namely a change in military retirement benefits.

Section 3: Application to Military Retirement Reform

The current military retirement system dates back to the 1940s when a common system was created for officers and enlisted personnel in all branches of the armed services. The current active duty military retirement system is a defined benefit plan that vests personnel at 20 YOS with a benefit based on a member's years of service and basic pay and that allows vested personnel to draw benefits immediately upon separation from service.

Numerous commissions, study groups, and analysts have critiqued the system.[1] The focus of the criticism has primarily been on the system that covers active duty members, though increasingly, the Reserve retirement system (which is related, but differs in key ways from the active one) has also been critiqued.[2] The main criticisms of the current system are that it is unfair to the vast majority of personnel who do not serve for 20 years (primarily enlisted personnel); it is a one-size-fits-all system that produces a one-size-fits-all experience mix that emphasizes youth and vigor over experience, regardless of community or occupation, even though some communities could benefit from longer careers; and it is inefficient because retirement benefits are not highly valued by most military personnel, who are young, yet such benefits are highly costly to provide. Regarding the last critique, the inefficiency occurs because the same retention could be achieved at less cost if the military provided more compensation up front in the form of current compensation. Another critique that has been raised by some is the lack of comparability with civilian retirement systems and specifically the lack of a defined contribution plan for military members.

While an analysis of the critiques and summary of the literature on military retirement reform is beyond the scope of this report, we note that most recommendations for reform involve several features:

- Earlier vesting of military personnel.
- Shifting deferred compensation toward current compensation.
- Adding a defined contribution plan.
- Flexible pays that can be targeted to achieve variable career lengths.

---

[1] For an analysis of the critiques of the AC retirement system and a summary of the various commissions and studies, see Warner (2008), Christian (2006), and Asch and Warner (1994).

[2] An analysis of the Reserve retirement reform, including critiques of the Reserve retirement system is presented in AHM (2013).

Recent proposals have therefore consistently recommended a package of changes that involve several components, and these components might be included in future proposals:

- A less generous defined benefit component.
- A defined contribution component that vests before YOS 20 and pays out benefits at older ages, e.g., starting at age 60.
- An increase in current compensation such as continuation pay to sustain the force size.
- An increase in current compensation in the form of a separation pay component that can be used to achieve the current retention profile or can be targeted to achieve variable career lengths.

Clearly, these components correspond to the most common recommendations listed above.

This section considers a specific reform proposal. The proposal is not one that is currently under consideration by DoD or one that we are recommending, though the proposal is similar in many respects to the proposals recommended by the 10th Quadrennial Review of Military Compensation, the 2006 Defense Advisory Committee on Military Compensation, and the 2000 Defense Science Board Task Force on Human Resources Strategy. We chose specific proposal features because they have the above four components and serve the purpose of demonstrating the capability we have developed to assess alternative proposals both in the steady state and in the transition to the steady state.

The specific details of the proposal we chose were selected to produce roughly the same retention as the current system in terms of the size of the active force and the experience mix. Any new military retirement system should, at a minimum, be able to yield the same force profile as under the current system. Thus, the retention profile that results from the proposal will, by design, be the same as the current retention profile. Consequently, the changes that occur in the steady state as well as in the transition to the steady state will be along other dimensions. For example, the proposal we consider results in a cost savings relative to the current system.

We note that while we selected specific details of the proposal for the purpose of our demonstration, we can accommodate a large array of different specifications that might be considered more desirable by policymakers. Again, our purpose is not to argue for a specific proposal, but to demonstrate a capability to simulate effects of retirement reform policies in the transition to the steady state and to simulate the effects of alternative implementation policies.

In this section, we show the effects of different implementation policies on retention and on cost. We first consider the case of full grandfathering and then the case of full grandfathering with choice. A key finding is that grandfathering with choice allows changes in retention and cost to be realized sooner than the case of full grandfathering without choice.

We begin with a description of the features of the proposal we analyze. We next show the effects of the proposal on retention and cost in the steady state and then consider the effects in transition to the steady state under different transition policies. We conclude with a brief summary.

Details of the Specific Retirement Reform Proposal We Consider

The retirement reform proposal we consider is a package of the four components indicated above.

Defined Benefit Component

The current military retirement benefit vests at 20 YOS in an immediate benefit that has the following formula:[3]

$$\text{Years of service} \times \text{highest-three years of basic pay} \times 0.025.$$

Because the current benefit allows vested members to draw benefits immediately when they separate, it pays retirement benefits not only in old age but also while they are pursuing their second career in the civilian sector. That is, if the typical member is age 42 when he or she retires from the military, the member receives benefits from age 42 to age 60—the second career—as well as after age 60.

The proposal we consider would continue to vest at 20 YOS but would reduce the benefit paid by reducing the multiplier from 0.025 to 0.01 and would allow members to draw benefits only when they reach age 60. That is, it would eliminate the second career payments of retirement benefits, but it would continue to vest at 20 YOS.

---

[3] There are actually three retirement systems currently in effect. The formula in the text is for the plan known as "High-Three" and is relevant for those who entered after 1980 and before 1986, as well as those who entered after 1986 and opted for the High-Three plan. For those entering service prior to 1980, the formula is based on last year of basic pay rather than highest-three years of basic pay, i.e., .025*YOS*final basic pay. For those entering after 1986, personnel have the choice at YOS 15 between High-Three or coverage of a plan known as "REDUX." The annuity formula under REDUX is less generous, (0.40 + 0.035*(YOS-20))*High-Three basic pay. Those who choose REDUX receive a $30,000 lump sum Career Status Bonus and are required to stay until YOS 20. The three plans have other differences such as differences in the cost of living adjustment.

Defined Contribution Component

Currently, military members have the opportunity to contribute to a defined contribution retirement plan that covers federal civil service workers, known as the Thrift Savings Plan, but DoD makes no contributions on behalf of the members (though each service branch can opt to do so).

Like the 10th Quadrennial Review of Military Compensation (QRMC), the proposal we consider would vest personnel who have completed 10 YOS in a defined contribution (DC) plan where the value of the benefit depends on the value of the DC fund when the member can begin drawing from the fund. We assume the member is eligible to begin drawing benefits at age 60. The value of the fund depends on the level and time pattern of contributions to the fund as well as the growth rate in the fund (which in turn depends on how the fund is invested). We assume that at each year of service, beginning at YOS 1 and ending at YOS 30, DoD automatically contributes 5 percent of a member's annual basic pay to the member's fund, and we assume that these contributions grow at a 5-percent real annual rate.

Increase in Current Compensation: Continuation Pay

The proposal's changes to the retirement system would reduce the value of retirement insofar as the defined benefit multiplier is reduced and would not include benefits during the second career phase of a member's career. While the addition of the defined contribution benefit offsets this reduction, we find (not shown) that without any other changes, retention would fall and the force size would decrease. Thus, to sustain retention, current compensation must increase. Different commissions, study groups, and analyses have considered alternative ways to increase current compensation. For example, the 10th QRMC recommended "gate pay," which is pay given to those who complete a specific career milestone, say 15 YOS, regardless of whether they stay or leave upon reaching the milestone.

For simplicity, we consider a continuation pay for Army enlisted personnel that is paid to those who reach 12 YOS. The continuation pay is a multiple of monthly basic pay. We assume that half the pay is paid at YOS 12 and the remaining amount is paid annually after the completion of the next 3 years. Thus, the pay is spread out over 4 years (YOS 12 to 15). For example, if a member's monthly basic pay were $3,200 at 12 YOS, the continuation pay would be equal to the multiplier × $3,200. If the multiplier were 3.99 (see next paragraph), the amount would be $12,768. Half of this amount ($6,384) would be paid in YOS 12, and the remaining half would be evenly spread over the following 3 years (or $2,128 per year).

The steady-state model includes an optimizer that allows us to compute the continuation pay multiplier that minimizes the difference between the retention profile under the reform proposal and the baseline profile under the current compensation system. We find that the optimal continuation pay multiplier is 3.99 times monthly basic pay. We use this multiplier when we model the transition to the new steady state. Because the multiplier is optimized to fit the baseline profile, it is not surprising that the retention profile under the reform proposal is nearly identical to the baseline profile. However, as we discuss shortly, personnel costs are not identical.

Increase in Current Compensation: Transition Pay

The final component of the reform package we consider is transition pay. Transition pay is provided to members who separate but is targeted to specific years of service. The formula is a multiple of annual basic pay:

$$\text{Multiplier} \times \text{Annual Basic Pay}.$$

The purpose of transition pay is to provide the services with force management flexibility to shape the force, especially in specific occupations or communities, as they see fit to meet their mission requirements. For example, personnel in occupations where a longer career is desired would not have transition pay available to them until a later YOS.

For the purpose of our demonstration, we assume that the services wish to replicate the current retention profile, and so we assume that transition pay is vested at 20 YOS so that those who separate with 20 or more YOS would receive this benefit. We make this assumption because at a minimum any reform proposal must ensure that the current retention profile is achievable. Therefore, the transition pay multiplier is optimized, together with the continuation pay multiplier, to replicate as closely as possible the baseline retention profile under the current compensation system. We find that the optimized transition pay multiplier is 2.92. It is useful to note that there is an interaction between the continuation pay and transition pay. Both induce higher retention among those with fewer than 20 YOS, with continuation pay focusing on inducing higher retention among those with fewer than 15 YOS. However, unlike continuation pay, transition pay also induces more separations, once members are eligible to receive it, beginning at YOS 20.

Steady-State Effects

It is useful to show the steady-state effects of the proposal to provide context for our results with respect to the transition to the steady state. Figure 3.1 shows the steady-state retention profile under the current compensation system and the steady-state profile under the specific proposal we are considering for Army enlisted personnel. The two profiles are virtually identical, and this is by design. As mentioned, we selected the continuation pay and transition pay multipliers to produce this result. Thus, the proposal produces almost no change in retention.

However, the proposal produces a cost savings in personnel costs in the steady state, measured as current compensation costs plus the retirement accrual costs, on the order of about $1.8 billion annually for the active Army enlisted force. The cost savings occur because the proposal reduces retirement benefits, a form of compensation that is more expensive to provide than it is valued by the typical enlisted member, and increases current compensation, a form of compensation that is more valued than retirement benefits by enlisted members.

The proposal has other effects, and our model can produce estimates of these effects although we do not focus on them in this report. For example, more personnel are vested in the new retirement system since the defined contribution plan vests at YOS 10, and so more personnel will be able to claim retirement benefits than under the current system. The proposal will also tend to reduce the present discounted value of compensation over the military career. That is, from the standpoint of lifetime benefits to the member over a career, enlisted members will be paid less. Thus, the proposal can achieve the same retention profile at less cost by reducing the present discounted value of compensation to military members. This finding implies that the current system is inefficient because the same force management outcome could be achieved at less cost. The finding also provides insight into why existing members and veterans oppose reform and why reform has not occurred. Addressing the inefficiency involves reducing military compensation, but this must be done in a way that leaves members as well off as they are under the current system so they will have equal willingness to stay in the military. We could design a reform proposal that holds the value of military compensation constant and so hold military members harmless in terms of changes in their lifetime compensation, but doing so would also reduce the cost savings. Thus, there are tradeoffs between the design features, the cost savings, and the change in the value of lifetime benefits to the member. We do not show these tradeoffs in this report but note that these are additional effects of the proposal to consider, and our model permits computation of these tradeoffs.

Figure 3.1
Steady-State Army Enlisted Retention Profile:
Current System Versus the Reform Proposal

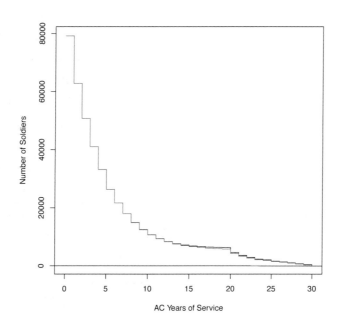

NOTE: The black line is the current system, and the red line is
the reform proposal.

The remainder of this section focuses on the transition to the steady state. We focus on the retention and cost effects of the proposals.

The Transition to the Steady State Under Alternative Implementation Strategies

We consider two transition policies: Full grandfathering of existing members and full grandfathering with choice to switch. Under the latter policy, existing members who would like to switch are permitted the opportunity to do so in the year the policy change occurs. That is, members are given a one-time opportunity to switch and the window of opportunity to switch lasts for one year.[4]

_____

[4] The model can accommodate longer windows of opportunity to switch to the new system, but this option adds considerable complication in terms of computation. We do not pursue this extension in this analysis.

We begin by considering the full grandfathering case. We show in this subsection that retention remains unchanged, by design, but it takes substantial time for new entrants, who are covered by the new system, to flow into the system and for existing members, covered by the existing system, to flow out. Consequently, it takes quite some time before the full cost savings of the change are realized. We also show that fully grandfathering members, but allowing existing members the option to switch to the new system also sustains retention. We find that not all existing members opt to switch but many do, especially more junior members, and this allows cost savings to be realized sooner than in the case with no option to switch.

Full Grandfathering Results

Figure 3.2 shows the retention survival curve by year of service as time elapses from when the policy change occurs. That is, the policy change occurs in year $t_a = 0$, and the figure shows the retention curves for $t_a = 0$ to $t_a = 30$. We find that retention is virtually unchanged by the policy, and so the curves are unchanged as time elapses. This occurs because of how we designed the policy change. The model optimizes continuation pay and transition pay multipliers to minimize the distance between the retention profile under the reform proposal and the baseline compensation system. Thus, Figure 3.2 is the counterpart of the steady-state retention profiles shown in Figure 3.1.

The results in Figure 3.2 belie the fact that changes actually occur as a result of the full-grandfathering implementation strategy because it includes both the retention of those who are grandfathered and those who are not (who comprise more and more of the younger portion of the force as successive new cohorts enter). When we consider the retention of each of these groups separately, we can observe the changes occurring over time.

Figures 3.3 and 3.4 show the retention survival curves of members grandfathered under the existing system while Figures 3.5 and 3.6 show the curves of new members who are not. Figures 3.3 and 3.5 show survival curves in two-dimensions for selected elapsed years while Figures 3.4 and 3.6 show survival curves in three-dimensions for all elapsed years, for $t = 0$ to $t = 30$. As time elapses since the policy occurs, more and more members are covered by the new system while fewer and fewer members are covered by the old system. Initially retention of junior members reflects retention behavior of those covered by the new system while retention of senior members reflects behavior of those covered by the existing system. As time elapses and grandfathered members flow out of the system, retention reflects more and more the behavior of those under the new system. After 30 years, the new system covers all members.

Figure 3.2
Retention Curves Under Retirement Reform with Grandfathering, by Time Elapsed Since the Policy Change

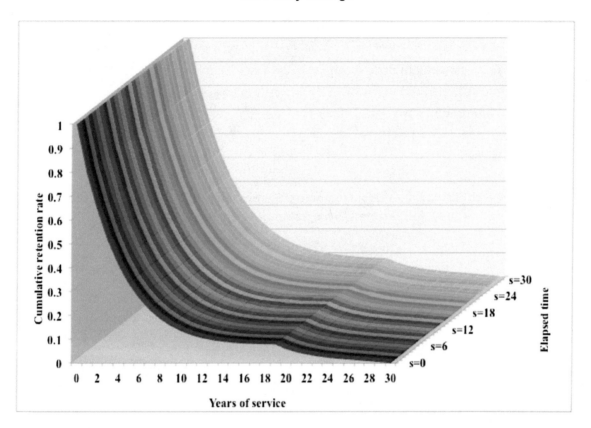

The effects on cost as time elapses are shown in Figures 3.7 and 3.8. Figure 3.7 shows the percentage change in total personnel costs as time elapses. Cost savings are realized gradually as retirement reform becomes more pervasive and more members are covered by the less costly retirement system. Figure 3.8 shows the percentage of total costs attributable to each component of compensation, by time elapsed, excluding regular military compensation.[5] In the early years of reform, costs are dominated by the defined benefit (DB) costs associated with those who are grandfathered, though it also includes the DB costs of those under the new plan. As time elapses, the costs of providing

---

[5] Regular military compensation (RMC) costs equals the sum of basic pay costs, the cost of the housing allowance and subsistence allowance, and the cost of providing allowances tax free. Total RMC costs across the force comprise the majority of total personnel costs, and total RMC costs do not change under retirement reform. However, their share of total costs (not shown) changes as the costs of the other components change as a result of retirement reform. We exclude RMC from the figure so we can highlight the costs that do change, namely retirement costs and the costs of the continuation pay and transition pay benefits.

contributions to the DC plan increase as more members are under the new system. At YOS 12, non-grandfathered members are eligible for the continuation pay (labeled as SRB cost in the figure), and at YOS 20, they are eligible for the transition pay. Thus, we observe continuation pay costs beginning at YOS 12 and transition pay costs beginning at YOS 20. By YOS 30, DB costs are a much smaller fraction of total costs than in the early years of implementation, covering only the less-generous DB plan offered under retirement reform. In addition, by YOS 30, transition pay and DC plan costs account for a larger fraction of total costs.

Figure 3.3
Retention Curves Under Retirement Reform for Grandfathered Members Only, for Selected Elapsed Years Since the Policy Change

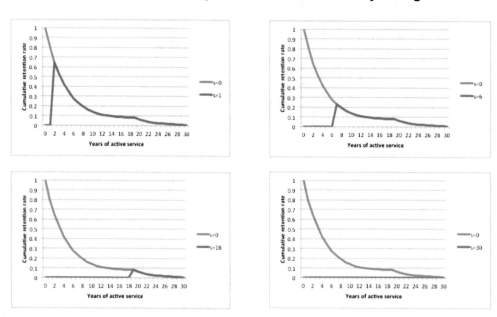

Figure 3.4
Retention Curves Under Retirement Reform for Grandfathered Members Only, by Time Elapsed Since the Policy Change

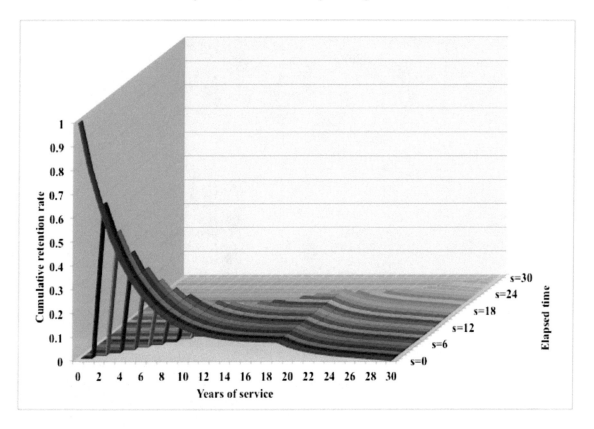

Consistent with the steady-state model, Figure 3.7 shows that the retirement reform proposal we consider yields a cost savings, on the order of 9 percent, when all members are under the new system. However, the path of cost savings is not even and continuous. The figure shows that the cost savings are greatest after 19 years have elapsed and is slightly less thereafter. Figure 3.8 shows why this is the case. Additional cost savings are slowly realized each year of implementation for the first 19 years. But beginning at 20 years, DoD incurs the cost associated with paying transition pay to those covered by the new plan. This additional cost stops the continuous decline in cost, and in fact, partially increases cost relative to the 19th year. After 30 years, when all members are under the current system, the steady-state cost savings of 9.0 percent are realized. These costs were 18 percent in period zero and 9 percent in period 30; hence, the cost savings of 9 percent (18 – 9 = 9).

Figure 3.5
Retention Curves Under Retirement Reform for Non-Grandfathered Members Only (e.g., New Entrants) for Selected Elapsed Years Since the Policy Change

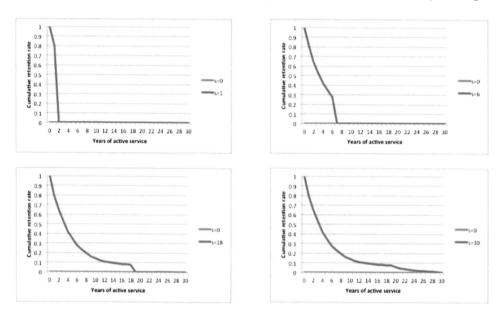

Results for Grandfathering with Choice to Switch to New System

An alternative strategy is to immediately convert existing members to the new system. In this case, cost savings would be realized immediately. As previously discussed, this has a strong political downside, namely that existing members would not be protected from the change and policymakers would gain the reputation of breaking faith with existing members. Thus, intermediate strategies are desirable.

Grandfathering with the option to switch is an intermediate strategy that has many desirable characteristics, as we show in this subsection. This strategy allows cost savings to be realized faster than in the case with no option to switch, while still keeping faith with existing members. Another desirable characteristic is that it is consistent with the concept of a volunteer force and market-based compensation system that is the foundation of the current manning approach to the U.S. military.

Figure 3.6
Retention Curves Under Retirement Reform for Non-Grandfathered Members Only (e.g., New Entrants), by Time Elapsed Since the Policy Change

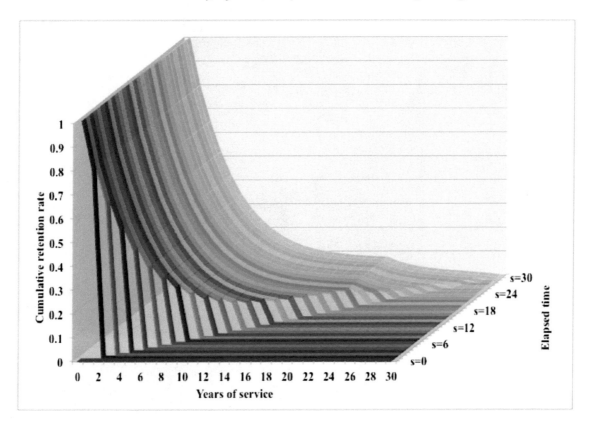

Figure 3.7
Percentage Change in Total Personnel Costs Under Retirement Reform (Full Grandfathering of Current Members) Relative to Baseline Costs, by Time in Years Elapsed Since the Policy Change

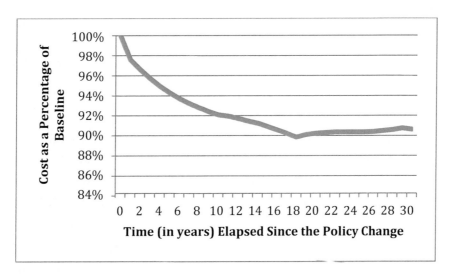

Figure 3.8
Compensation Component Costs as a Percentage
of Total Personnel Costs Under Retirement Reform,
by Time Elapsed Since the Policy Change

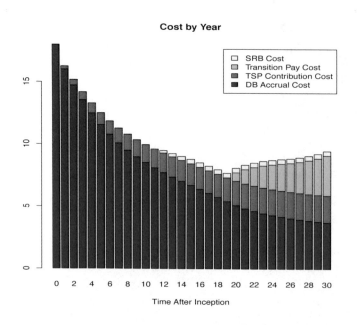

NOTE: The chart stacks the components of compensation cost.
The stacks at each time *s* do not sum to 100 percent because we
exclude regular military compensation costs in each year. That is,
RMC is the missing category of costs.

Figure 3.9 shows the retention effect of retirement reform over time under this
implementation strategy. As before, retention is unchanged because of the specific
design features we chose for the retirement reform package. Thus, Figure 3.9 is the
counterpart to Figures 3.1 and 3.2. However, as before, the lack of change in overall
retention belies changes in terms of coverage by the new plan; retention at a given year of
service at a point of time will be a mix of retention of those covered and not covered by
retirement reform. Because some grandfathered members can switch, the retention
profile of those under the new system at a given year of service will include those
automatically covered by the new policy (e.g., new entrants) and those grandfathered
members who switched to the new policy.

Figure 3.10 shows the switching behavior under this implementation strategy by
cohort, defined as the member's year of service at the time of the policy change. The
figure shows the percentage of members at each year of service who participate in
retirement reform, either because they switched to the new system or because they were
automatically placed under the new system. The figure illustrates two results. First,
more senior cohorts are less likely to choose to switch to the new retirement system, and

second, for a given cohort the fraction in the new plan declines over the career because of the pattern of retention over the career of those who opt to switch in that cohort.

Figure 3.9
Retention Curves Under Retirement Reform with Grandfathering and Switching, by Time Elapsed Since the Policy Change

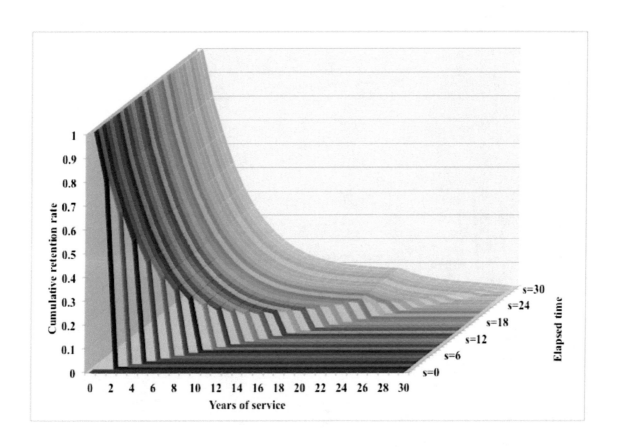

Figure 3.10
Percentage of Personnel at Each YOS Who Participate in Retirement Reform by Cohort
(Defined by YOS at the Time of the Policy Change)

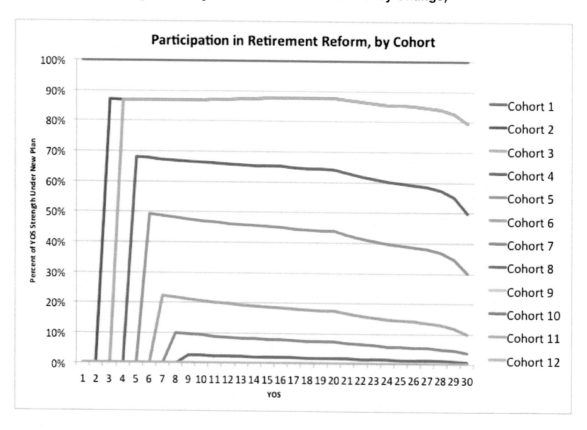

NOTE: Cohorts 9-12 are not seen in the figure because the participation rate is 0 percent.

More specifically, Figure 3.10 shows that 100 percent of members in cohort 1 participate in retirement reform, regardless of year of service. The reason is that the cohort of new entrants is automatically covered by the new system. We find that 87 percent of cohorts 2 and 3 opt into the retirement reform, and this percentage remains fairly constant over the following years of service for these cohorts. That is, we find that 87 percent of junior enlisted members who were in YOS 2-3 at the time of the policy change opt to switch to the new system because the value of their military career (the V_a used to compute the cumulative probabilities in equation 16) is greater by doing so. Also, the retention in following years of those who opt in is essentially the same as those who do not opt in, so in any future years, about 87 percent of the remaining members of the cohort are those who opt in. Fewer members in YOS 4-5 choose to switch, 68 percent and 49 percent, respectively. By YOS 6, only 22 percent opt to switch, and by YOS 10, no members opt to switch. Further, in these cohorts, retention is somewhat lower among those who opt in, so those who opt ins are a declining percentage of the remaining members of the cohort in future years.

Thus, some, but not all, members in YOS 2-9 opt to switch. Those who switch have different tastes for service than those who do not switch. The provision of continuation pay at YOS 12 increases the value of staying in the AC for junior personnel. Continuation pay is sufficiently compelling for 87 percent of junior enlisted members who were in YOS 2-3 at the time of the policy change. However, among cohort 9, only 3 percent opt to switch. The percentage that switch in cohort 9 is related to the member's taste for service in that cohort. It is true that those in cohort 9 can receive the bonus in three more years of service, so the present value of the bonus is higher for them than members with fewer years of service. But it is also true that the stream of payouts under the baseline system, including the second-career retirement payouts, is closer and less discounted than it is for more junior cohorts. In the model, however, the discount factor applied to future payouts is the same for everyone in the cohort, so the fact that some switch and others do not is not a result of discounting. Instead, it is a result of different tastes for service among those in cohort 9. Those with relatively lower tastes for service are more likely to have shorter careers, and the higher near-term payout under continuation pay is more attractive than the more deferred, but higher compensation under the baseline system. Among those in cohort 10, discounting again comes to the fore; the new system is always less valuable to senior members, regardless of their taste for military service. Thus, none of the more senior personnel at the time of the policy opts to switch. As a further point, grandfathered members who switch to the new system and stay beyond YOS 12 will also face lower payouts in their remaining years, but these members did receive the continuation pay at YOS 12. At the time they made the decision to switch to the new system, say at YOS 9, they had the foresight to recognize both the continuation pay at YOS 12 and the lower payouts in later years and chose rationally based on this information. Their behavior in the model is time consistent and without regret.

The results of this switching behavior in terms of retention behavior are shown in Figures 3.11-3.14. Figures 3.11 and 3.12 show the retention survival curves for those under the current plan (i.e., grandfathered members who opt not to switch to the new system), and Figures 3.13 and 3.14 show the retention survival curves for those under the new plan (i.e., automatically covered members and grandfathered members who switched). As before, the two-dimensional figures (Figures 3.11 and 3.13) show survival curves for selected years, while the three-dimensional ones (Figures 3.12 and 3.13) show them for all years. In year zero (the year of the policy change), all existing members are under the current plan, and no one is under the new plan. However, in year 1 (1 year after the policy change), many of the junior personnel under the current system switch over, and so the force profile for those under the current plan has relatively few junior personnel. As time elapses, those who opted to stay under the existing plan get older and more senior. After 30 years of elapsed time, all existing members have separated.

Because some existing members choose to switch, more cost savings are realized early in the implementation of the new policy. This is seen by contrasting the pattern of cost savings under grandfathering without switching and grandfathering with switching. Figure 3.15 replicates the results in Figure 3.7 and also shows the evolution of the percentage change in total personnel costs over time when members are permitted to switch to the new system. Unlike grandfathering without switching, in which costs savings are realized slowly in the early years, immediate cost savings are realized in the first year of implementation. The reason is seen in Figure 3.16, showing the percentage of total cost associated with each component of cost at each year of service. Because many existing members switch in the first year of the policy, retirement accrual costs fall dramatically in the first year in Figure 3.16. Cost savings increase gradually after the first year through year 13. Beginning in year 14, members who switched when they were junior members are beginning to reach 20 YOS and become eligible for transition pay. The addition of the cost of transition pay slows the increase in cost savings as time elapses. The steady-state cost savings are achieved after 30 years.

Figure 3.11
Retention Curves for Those Under the Current Plan for Selected Elapsed Times Since the Policy Change

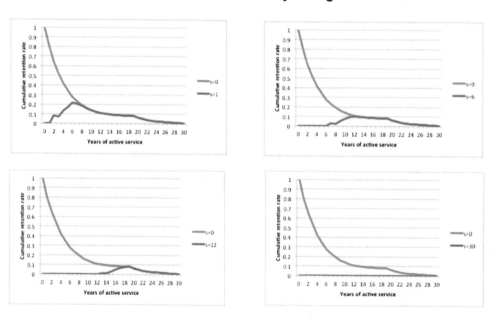

Figure 3.12
Retention Curves for Those Under the Current Plan, by Time Elapsed Since the Policy Change

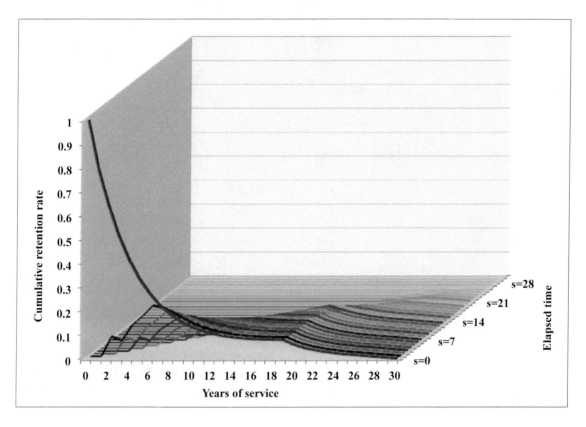

Figure 3.13
Retention Curves for Those Under the New Plan for Selected Elapsed Times Since the Policy Change

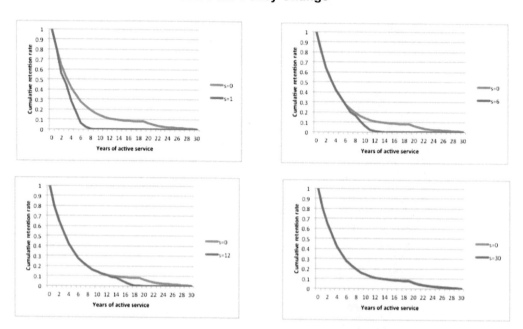

Figure 3.14
Retention Curves for Those Under the New Plan, by Time Elapsed Since the Policy Change

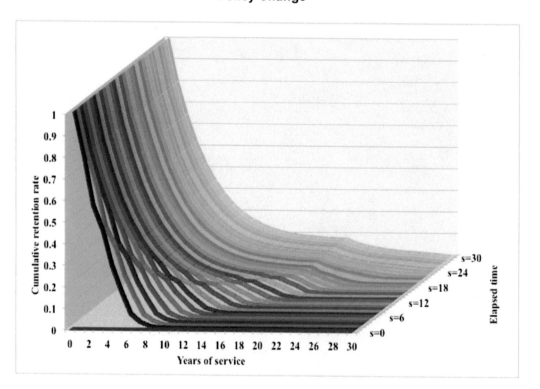

Figure 3.15
Percentage Change in Total Personnel Costs Under Retirement Reform with Grandfathering and Switching Relative to Baseline Costs, by Time Elapsed Since the Policy Change

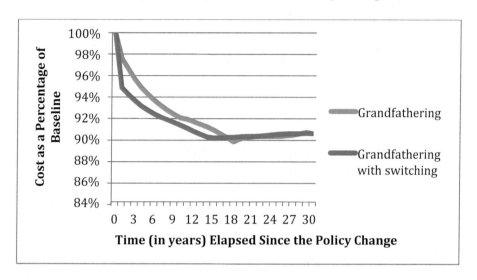

Figure 3.16
Compensation Component Costs as a Percentage of Total Personnel Costs Under Retirement Reform with Grandfathering and Switching, by Time Elapsed Since the Policy Change

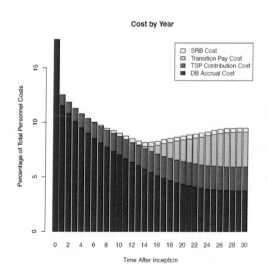

NOTE: The chart stacks the components of compensation cost. The stacks at each time s do not sum to 100 percent because we exclude regular military compensation costs in each year. That is, RMC is the missing category of costs.

Summing Up

This section considered a proposal that has the four elements generally recommended by recent study groups and commissions as elements of retirement reform, though the specifics of our proposal differ from past proposals. The reform proposal we assess is designed to hold AC retention constant. We considered two implementation strategies. The first—full grandfathering of existing members—is one commonly recommended by policymakers. Indeed, as the quote in the introduction makes clear, Secretary Panetta has promised members currently serving in the military that they would be fully grandfathered should retirement reform be adopted. The second is a variation of the first and involves fully grandfathering existing members but also allowing them the choice to switch to the new system during a one-year period. Like the first strategy, the second one is consistent with keeping faith with existing members in terms of providing the retirement benefit they were promised when they joined the military, but it allows members greater choice.

We find, as expected, that fully grandfathering existing members results in a gradual conversion of members from the old system to the new system as new members covered by the new system enter and age, and existing members covered by the old system age and exit. Consequently, the cost savings associated with retirement reform are gradually realized over time. In contrast, we find that many members opt to switch to the new system when given the option to do so. For the most part, those who opt to switch are more junior at the time of the policy change. Because some members switch, more of the cost savings associated with retirement reform are realized early on. That is, while the full cost savings of retirement reform are realized only when the new steady state is reached, much of the saving is realized in the first year after members have a chance to switch.

The purpose of the analysis is not to recommend a specific proposal but to demonstrate our new capability to consider the transition to the new steady state. The analysis shows that while the same steady state is realized, the transition policy can affect the time path of change. The timing of change can influence whether a policy such as retirement reform is adopted.

Section 4: Concluding Thoughts

The research summarized in this report extends the dynamic retention model to simulate the transition to the steady state. This extension provides researchers with the ability to assess the effects of workforce management policies both in the steady state and in the transition to the steady state as well as to assess the effects of alternative implementation strategies. Consequently, it greatly enhances the capability of research to support evidence-based decisionmaking with respect to workforce management. It also helps policymakers better understand workforce dynamics and how they respond to policy change.

The transition modeling presented in this report has major advantages over other approaches that are commonly used to assess the retention effects of different policies in the transition period. The DRM models behavior with a dynamic program, and the model is structured so all relevant information about the individual's history is summed up in the state variables, namely, active years of service, Reserve years, total years, and status (AC, RC, civilian). Further, the model allows individuals to differ in their tastes for active and Reserve service, and the model assumes the individual is buffeted by shocks that can affect the decision. The model's structure leads to an expression for the probability in each year that the individual will take a certain action given the current state, e.g., stay in the AC, leave and become a pure civilian, or leave and join the Reserve. At the micro level, this structure is a first-order Markov process; the transition from the current state at time t to a new state at time $t + 1$ is Markovian. But the DRM is quite different from other commonly used implementations of Markov transitions.

For example, one common approach is called inventory projection modeling. Under this approach, analysts might use historical transition probabilities to project the evolution of a personnel inventory. Historical rates, however, are inappropriate for the transition probabilities under a new policy; one cannot assume that the transition probabilities will remain the same as they have been. Instead, the modeler might estimate new steady-state transition probabilities, use these to replace the baseline transition probabilities, and run the inventory projection model forward from its state at the time the new policy went into effect to develop retention profiles in the transition period and the eventual new steady state. While this approach will provide estimates of retention in the transition period, the approach is severely limited for two reasons. First, it cannot incorporate behavioral changes among the incumbent population that may occur during the transition to a new steady state. That is, the Markov approach assumes the behavior in the transition period is the same as the new steady state and does not allow different behavior in the transition. Thus, for example, the Markov approach cannot accommodate an implementation strategy where grandfathered members have the option to switch to the new system because it cannot accommodate any behavioral effects. Similarly, it cannot easily accommodate temporary policies that are intended to influence behavior only during the transition period, such as a short-term separation pay program. As shown in Section 3, the DRM transition modeling can accommodate behavioral change in the transition.

Second, the estimates of the new steady-state transition probabilities may be time-inconsistent or worse. Another common approach to model the transition effects of compensation policies is to apply retention elasticity estimates, usually garnered from other studies. An elasticity estimate is an estimate of the percentage change in retention due to a percentage change in compensation. Econometric studies of retention often use data on the retention decisions of a workforce and estimate the effects of pay on retention holding constant other factors that could affect retention, such as demographic and job characteristics. The pay variable might be constructed as current military pay relative to current civilian pay, or perhaps military pay over 4 years relative to civilian pay over 4 years. This approach ignores future compensation, differences in taste among members, and differences in history (who stayed/who left) affecting the posterior distribution of taste for those making the retention decision. Further, the approach does not address the issue of time consistency of behavior.

In contrast, the DRM approach is logically consistent with rational behavior over time and handles decisionmaking under uncertainty in an attractive way by allowing people to change their decisions over time as uncertainty is resolved and new information becomes available to them. By extending the DRM to the transition period, we are now able to extend the advantages of the DRM to the transition period as well. As mentioned at the outset, this allows the analysis to consider policies aimed at facilitating the transition to a new force size and shape, policies of a long-term nature such as changes to the basic pay table or the retirement benefit structure, and the total and year-by-year cost of a policy change that might involve both short- and long-term features.

This study focused on specific examples related to separation pay and retirement reform, but the capability we developed can be used or further extended to consider a wide array of other policies, including pay freezes, bonuses, targeted special and incentive pays, changes in promotion speed as well or other promotion policy changes, and other retirement reform proposals. A related RAND project is currently using the capability to consider retirement reform proposals under consideration by the Department of Defense as part of its comprehensive review of military compensation.

While our analysis is applied to active component military personnel, it can be applied to other workforces including Reserve component personnel, federal civil service personnel, state and local government employees, and even private-sector workforces. Applying the methodology to other workforces will require econometric estimates of their underlying DRM parameters.

A related aspect of this point has to do with the analysis and management of multiple cohorts of a workforce. For example, special operations forces have become increasingly important and their number has increased. The approach we have used to analyze the overall force estimates the model on the retention behavior of the 1990/91 entering cohorts tracked to 2010. This approach may be problematic for special operators because of their relatively small number in the 1990/91 cohort and changes in their pay and personnel policy since that time. A possible alternative approach is to analyze entry cohorts from, say, 1990 to 2004, and—taking advantage of ideas developed here—specify the model to allow for anticipatory retention behavior in response to changes in special pays, such as the critical skills retention bonus introduced in 2007, which among others targeted special operators with 19-23 YOS.

Analyzing multiple entry cohorts might also enable tests of whether the taste distribution parameters of entering cohorts have changed from cohort to cohort. Changes could result from changing expectations about deployment, compensation, military training, educational benefits, and so forth. Further, it might be possible to recover the distribution of taste for the military in the youth population (Hosek et al., 2004, do so with a calibrated model).

The model has many similarities to dynamic programming research on labor supply, saving, and retirement (e.g., van der Klaauw and Wolpin, 2008; Gustman and Steinmeier, 2005), but the lack of longitudinal data on military spouse employment, post-service employment, wealth, health, and retirement are major impediments to extending the model to look at behavioral responses along these margins in addition to retention.

This study also focused on permanent policy changes that aimed to change compensation in the steady state. However, the capability could also be used to consider temporary policies specifically designed to affect the transition to the steady state, such as separation pay that is available to members for only a limited amount of time. Another related RAND project is using the modeling approach we develop here to consider the effects of using temporary separation pay to facilitate the Army drawdown.

Finally, the algorithms developed here to simulate the policy changes presented in the examples can be viewed as a step toward developing the algorithms needed to extend the capability of DRM to be estimated on historical data over multiple cohorts where compensation policy varies by cohort. The approach to estimating the DRM so far has assumed a steady state, but there have been changes in special and incentive pays and in the level and structure of military compensation relative to civilian compensation. These changes, some of which apply for a fixed period of time, affect certain groups, or allow a menu of choices, can in principle be represented through a more complex likelihood function than used for the steady-state estimation. The development of this enhanced estimation capability awaits future research.

Appendix: Additional Model Results Related to Retention Effects

This appendix focuses on how active retention (Pr(A) in the main text) changes when different model parameters change and how the responsiveness of active retention to changes in the value of an active career changes when the variance of the taste distribution changes.

Effect of Changes in the Variance of the Shock on Active Retention

We start by considering how Pr(A) in equation (7) in the main text changes when the variance of the shocks change. The AC shock and the Reserve-civilian nest shock have the same scale parameter, κ, and higher values of κ imply a high shock variance; the variance is $\pi^2 \kappa^2 / 6$. We show that the effect of a higher shock variance on Pr (A) is ambiguous. We then consider a change in the scale parameter λ for the error of the within-nest choices Reserve and civilian.

Active is chosen if the payoff to active exceeds the payoff to the Reserve-civilian nest:

$V_a + \varepsilon_a > Max(V_c + \omega_c, V_r + \omega_r) + \upsilon_{rc}$, where $\omega_c, \omega_r \sim ExtremeValue(0, \lambda)$ and $\upsilon_{rc} \sim ExtremeValue(0, \tau)$.

That is, the within-nest choice-specific errors are distributed extreme value type 1, also known as a Gumbel distribution, with a shape parameter of zero and a scale parameter of λ. Further, it is known that the distribution of the maximum expression is also extreme value:

$$Max(V_c + \omega_c, V_r + \omega_r) \sim ExtremeValue(\lambda \, Ln[e^{\frac{V_c}{\lambda}} + e^{\frac{V_r}{\lambda}}, \lambda].$$

We now rewrite the right-hand side of the condition for choosing active as follows:

$$\lambda \, Log\left[e^{\frac{V_c}{\lambda}} + e^{\frac{V_r}{\lambda}}\right] + \omega'_{rc} + \upsilon_{rc} \text{ where}$$

$$\omega'_{rc} = Max(V_c + \omega_c, V_r + \omega_r) - \lambda \, Log\left[e^{\frac{V_c}{\lambda}} + e^{\frac{V_r}{\lambda}}\right].$$

$$\text{hence } \omega'_{rc} \sim ExtremeValue[0, \lambda]$$

Define $\varepsilon_{rc} = \omega'_{rc} + v_{rc}$. The distributions of ω'_{rc} and v_{rc} are both extreme value, with a location parameter of zero, and since they are independent, their variance of their sum is the sum of their variances, $\pi^2(\lambda^2 + \tau^2)/6$. For brevity, define κ such that $\kappa^2 = \lambda^2 + \tau^2$ so $\kappa = \sqrt{\lambda^2 + \tau^2}$. Thus, $\varepsilon_{rc} \sim ExtremeValue[0, \kappa]$. We want ε_a to have the same location and scale parameters, $\varepsilon_a \sim ExtremeValue[0, \kappa]$.

Returning to the choice condition and using the above expressions, we have

$$V_a + \varepsilon_a > \lambda \, Log\left[e^{\frac{V_c}{\lambda}} + e^{\frac{V_r}{\lambda}}\right] + \varepsilon_{rc}$$

$$\varepsilon_{rc} - \varepsilon_a < V_a - \lambda \, Log\left[e^{\frac{V_c}{\lambda}} + e^{\frac{V_r}{\lambda}}\right]$$

$$\varepsilon < V_a - \lambda \, Log\left[e^{\frac{V_c}{\lambda}} + e^{\frac{V_r}{\lambda}}\right] \text{ where } \varepsilon = \varepsilon_{rc} - \varepsilon_a.$$

The difference in two extreme value type 1 random variables has a logistic distribution; if $X, Y \sim ExtremeValue[\alpha, \beta]$, then $X - Y \sim Logistic[0, \beta]$. It follows that $\varepsilon \sim Logistic[0, \kappa]$. The functional form of the logistic distribution is

$$Pr(\varepsilon < \xi) = \frac{1}{1 + e^{-\frac{\xi}{\kappa}}}.$$

Using the logistic distribution, the probability of choosing active is

$$Pr\left(\varepsilon < V_a - \lambda \, Log\left[e^{\frac{V_c}{\lambda}} + e^{\frac{V_r}{\lambda}}\right]\right) = \frac{1}{1 + e^{-\frac{V_a - \lambda \, Log\left[e^{\frac{V_c}{\lambda}} + e^{\frac{V_r}{\lambda}}\right]}{\kappa}}}.$$

The right-hand side can be rearranged to

$$\frac{e^{\frac{V_a}{\kappa}}}{e^{\frac{V_a}{\kappa}} + e^{\frac{\lambda \, Log\left[e^{\frac{V_c}{\lambda}} + e^{\frac{V_r}{\lambda}}\right]}{\kappa}}}.$$

Further, using the fact that $e^{b \, Lna} = a^b$, this becomes the familiar form:

$$\frac{e^{\frac{V_a}{\kappa}}}{e^{\frac{V_a}{\kappa}} + (e^{\frac{V_c}{\lambda}} + e^{\frac{V_r}{\lambda}})^{\frac{\lambda}{\kappa}}}.$$

Taking the derivative of Pr (A) with respect to κ gives

$$\Pr(A)\,(1 - \Pr(A))\; \frac{-\left(V_a - \lambda \operatorname{Log}\left[e^{\frac{V_c}{\lambda}} + e^{\frac{V_r}{\lambda}}\right]\right)}{\kappa^2}.$$

The derivative is positive only if $V_a - \lambda \operatorname{Log}\left[e^{\frac{V_c}{\lambda}} + e^{\frac{V_r}{\lambda}}\right] < 0$. Looking back to the probability of choosing active, $\Pr\left(\varepsilon < V_a - \lambda \operatorname{Log}\left[e^{\frac{V_c}{\lambda}} + e^{\frac{V_r}{\lambda}}\right]\right)$, it follows that this probability increases with κ only if the right-hand side of the inequality is negative. Since the logistic distribution is symmetric and the location parameter of ε is zero, this means that an increase in κ increases Pr (A) whenever the value of $V_a - \lambda \operatorname{Log}\left[e^{\frac{V_c}{\lambda}} + e^{\frac{V_r}{\lambda}}\right]$ is in the negative range. A graphic illustrates why this is so. For instance, reading from Figure A.1, if the value of $V_a - \lambda \operatorname{Log}\left[e^{\frac{V_c}{\lambda}} + e^{\frac{V_r}{\lambda}}\right]$ were -20, then an increase in κ would increase the probability of choosing active from about 0.24 to 0.30.

Figure A.1
Distribution of ε

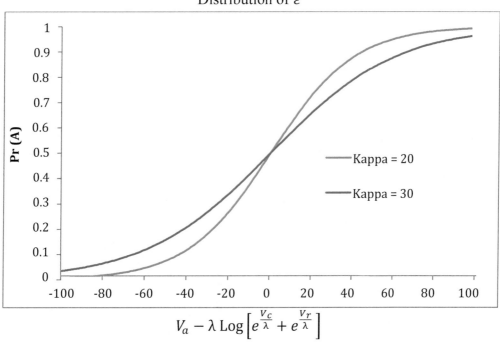

We next ask how an increase in λ affects the probability of staying in the AC. The variance of the within-nest shocks ω_c and ω_r is $\pi^2 \lambda/6$; so higher values of λ imply a higher within-nest shock variance. The derivative of $\Pr(A)$ with respect to λ is

$$\Pr(A)(1 - \Pr(A))(\frac{e^{\frac{V_c}{\lambda}}V_c + e^{\frac{V_r}{\lambda}}V_r}{e^{\frac{V_c}{\lambda}} + e^{\frac{V_r}{\lambda}}} - \lambda \operatorname{Log}[e^{\frac{V_c}{\lambda}} + e^{\frac{V_r}{\lambda}}])/(\kappa\lambda).$$

This derivative is positive whenever the expected value of the payoffs in the Reserve-civilian nest is greater than the mode of the distribution of the best draw from the nest. The expected payoff in the nest is $\Pr(V_c > V_r) \, V_c + \Pr(V_r > V_c)V_r$, or

$$\frac{e^{\frac{V_c}{\lambda}}}{e^{\frac{V_c}{\lambda}} + e^{\frac{V_r}{\lambda}}}V_c + \frac{e^{\frac{V_r}{\lambda}}}{e^{\frac{V_c}{\lambda}} + e^{\frac{V_r}{\lambda}}}V_r.$$

Whenever the expected payoff in the nest is greater than the mode of $\operatorname{Max}(V_c + \omega_c, V_r + \omega_r)$, an increase in λ decreases the probability of this relatively high draw and increases the probability that a given value of V_a exceeds the draw. This leads to an increase in $\Pr(A)$.

Effect of Changes in the Variance of the Taste Distribution on the Responsiveness of Active Retention to Changes in V_a

We next consider the responsiveness of the retention rate to an increase in the value of the military career and ask whether the responsiveness increases, or decreases, as the variance of taste increases. If taste variance is higher, when might there be reason to expect the responsiveness of retention to be, say, lower? Our analysis shows that the answer is in general ambiguous, but we offer an example consisting of two cases—of sets of parameters—consistent with service members at an early stage of their military career or at a later stage, though before reaching 20 YOS and retirement eligibility. The example indicates that among junior service members an increase in taste variance seems likely to decrease the responsiveness of the retention rate to an increase in the value of the military career at that point, whereas among more senior service members the reverse seems likely. In other words, the effect of an increase in taste variance on the responsiveness of retention to the value of the military career depends on the point of evaluation. To give a more specific introduction to the analysis and example, it is important to recognize that retention is a selective process: Personnel with a higher taste for military service are more likely to remain in the military, and this causes the taste distribution of the retained population to evolve from its initial value at the outset of service. The posterior mean taste increases, and the posterior variance of taste decreases. As a result, the example assumes that the mean taste of senior personnel is higher than that of junior personnel, and the variance (or standard deviation) of taste is lower for senior personnel. Another element that changes is the value of the military career, which is considerably higher as the member approaches retirement eligibility than it is in the early part of the career. These factors—higher mean taste, lower variance of taste, and higher value of the military career—work together to make an *increase* in taste variance likely to increase the responsiveness of retention to an increase in the value of the military career among senior personnel, and yet to decrease this responsiveness among junior personnel. The mechanism, or intuition, behind this comes from the fact that senior personnel are likely to be further out on the taste distribution relative to its senior mean, and in that range, an increase in variance causes an increase in the density of realizations more than a standard deviation from the mean. In turn, this increase in density drives the increase in retention responsiveness to the value of the military career.

Our model assumes that the ex ante (beginning of career) taste distribution is normal. The standard normal density is

$$\frac{1}{\sqrt{2\pi}\,\sigma}\, e^{\frac{-(x-\mu)^2}{2\sigma^2}},$$

and its derivative with respect to σ is

$$\frac{1}{\sqrt{2\pi}\,\sigma^4}\, e^{\frac{-(x-\mu)^2}{2\sigma^2}}\left((x-\mu)^2 - \sigma^2\right).$$

It follows that the derivative is positive when $|x - \mu| > \sigma$, that is, when $x - \mu$ is greater than $+\sigma$ or less than $-\sigma$, and negative otherwise. Increasing the variance decreases the normal density between $-\sigma$ and $+\sigma$ and increases the density in the tails beyond $\pm\sigma$.

A notationally simplified but accurate version of our model assumes that at a given year of service the value of the AC career, w, is the same for everyone apart from taste, which is an additive term. The individual's retention probability is

$$r = \frac{e^{\gamma+w}}{1+e^{\gamma+w}}.$$

The derivative of this probability with respect to w is $r(1 - r)$, which is always positive. As mentioned, the taste distribution will evolve during the career because of selective retention. In particular, the posterior mean taste will increase, and the variance of taste will decrease. For simplicity, we continue to assume that taste is normally distributed during the career, though in our example below the mean and variance change for senior versus junior personnel. The average retention rate may be written

$$R(w, \sigma) = \int_{-\infty}^{\infty} \frac{e^{\gamma + w}}{1 + e^{\gamma + w}} \frac{1}{\sqrt{2\pi}\,\sigma} e^{\frac{-(x-\mu)^2}{2\sigma^2}} d\gamma.$$

The change in the average retention rate with respect to w is

$$\frac{\partial R(w, \sigma)}{\partial w} = \int_{-\infty}^{\infty} \frac{e^{\gamma + w}}{(1 + e^{\gamma + w})^2} \frac{1}{\sqrt{2\pi}\,\sigma} e^{\frac{-(x-\mu)^2}{2\sigma^2}} d\gamma.$$

The change in this derivative with respect to the standard deviation of taste is the quantity we are interested in:

$$\frac{\partial^2 R(w, \sigma)}{\partial w\, \partial \sigma} = \int_{-\infty}^{\infty} \frac{e^{\gamma + w}}{(1 + e^{\gamma + w})^2} \frac{1}{\sqrt{2\pi}\,\sigma^4} e^{\frac{-(x-\mu)^2}{2\sigma^2}} ((x - \mu)^2 - \sigma^2) d\gamma.$$

The normal density does not have a closed form for integration but the integral can be numerically approximated. We have done this for two cases, Case 1 and Case 2. For each case, we set starting values and then change the standard deviation of taste to see how the responsiveness of retention to w changes. In Case 1, we have $\{w = 1, \mu = -1, \sigma = 1\}$. These values produce an average retention rate of 52 percent, which is in the range of first-term reenlistment for the Army, Navy, and Air Force. In examining Case 1, we hold w and μ at their set values and vary the value of σ, and similarly for Case 2. In Case 2, we have $\{w = 2.25, \mu = 0, \sigma = .7\}$. Compared with Case 1, Case 2 is more like the senior career. The value of the career in Case 2 is more than twice as high as in Case 1, as is consistent with a high expected value of retirement benefits as the member approaches 20 YOS. Also, the variance of taste is smaller in Case 2, which is consistent with higher retention over the military career of those with higher taste for the military. The taste variance in Case 1 is 1 and in Case 2 is 0.49.

The results of the evaluations are in Table A.1. In Case 1, an increase in taste variance, e.g., an increase in the standard deviation of taste from 1.0 to 1.1, decreases retention responsiveness to the value of the AC career, and this pattern is true over the entire range of standard deviation shown. In Case 2, the reverse is true. Specifically, an increase in the standard deviation of taste from 0.7 to 1.1 increases retention responsiveness to the value of the AC career, and again this result holds over the entire range of standard deviations shown. The results in Table A.1 illustrate our point that an increase in taste variance can increase or decrease the responsiveness of retention to the value of the career. As a side point, also note that in both cases, an increase in taste variance slightly decreases the average retention rate.

Table A.1
Retention Responsiveness to the Value of the AC Career at Different Values of Taste Variance

	Case 1		Case 2	
Taste s.d. (σ)	R	dR/dw	R	dR/dw
0.7	0.522	0.417	0.897	0.116
0.8	0.522	0.401	0.892	0.122
0.9	0.521	0.385	0.887	0.127
1.0	0.521	0.370	0.881	0.133
1.1	0.520	0.355	0.875	0.138

NOTE: s.d. is standard deviation. R is the average retention rate given by $R(w, \sigma)$ and dR/dw is the derivative of the average retention rate with respect to the value of an AC career, w.

We emphasize that this is only an illustrative example. The actual calculation is more complex because the posterior distribution of taste among those staying in the AC is not likely to be a symmetrical bell-shaped distribution, whereas the example assumes a normal distribution for both junior and senior members. Still, the example suggests that an increase in taste variance will decrease the retention responsiveness to an increase in w among junior members and can have the opposite effect among senior members.

Bibliography

Aguirregabiria, Victor, and Pedro Mira, "Dynamic Discrete Choice Structural Models: A Survey," *Journal of Econometrics*, Vol. 156, No. 1, 2010, pp. 38-67.

Asch, Beth J., James Hosek, and Michael G. Mattock [AHM], *A Policy Analysis of Reserve Retirement Reform*, Santa Monica, Calif.: RAND Corporation, MG-378-OSD, 2013. As of March 27, 2013:
http://www.rand.org/pubs/monographs/MG378.html

Asch, Beth J., James Hosek, Michael G. Mattock, and Christina Panis, *Assessing Compensation Reform: Research in Support of the 10th Quadrennial Review of Military Compensation*, Santa Monica, Calif.: RAND Corporation, MG-764-OSD, 2008. As of March 27, 2013:
http://www.rand.org/pubs/monographs/MG764.html

Asch, Beth J., Richard Johnson, and John T. Warner, *Reforming the Military Retirement System*, MR-748-OSD, Santa Monica, Calif.: RAND Corporation, 1998. As of March 27, 2013:
http://www.rand.org/pubs/monograph_reports/MR748.html

Asch, Beth J., and John T. Warner, *A Theory of Military Compensation and Personnel Policy*, MR-439-OSD, Santa Monica, Calif.: RAND Corporation, 1994. As of March 27, 2013:
http://www.rand.org/pubs/monograph_reports/MR439.html

Asch, Beth J., and John T. Warner, *Separation and Retirement Incentives in the Federal Civil Service: A Comparison of the Federal Employees Retirement System and the Civil Service Retirement System*, Santa Monica, Calif.: RAND Corporation, MR-986-OSD, 1999. As of March 27, 2013:
http://www.rand.org/pubs/monograph_reports/MR986.html

Asch, Beth, and John Warner, " A Theory of Compensation and Personnel Policy in a Hierarchical Organization with Application to the U.S. Military," *Journal of Labor Economics*, Vol. 19, 2001, pp. 523-562.

Bajari, Patrick, C. Lanier Benkard, and Jonathan Levin, "Estimating Dynamic Models of Imperfect Competition," *Econometrica*, Vol. 75, No. 5, September 2007, pp. 1331-1370.

Ben-Akiva, Moshe, and Steven Lerman, *Discrete Choice Analysis: Theory and Application to Travel Demand,* Cambridge, Mass.: MIT Press, 1985.

Borkovsky, Ron, Ulrich Doraszelski, and Yaroslav Kryukov, "A Dynamic Quality Ladder Model with Entry and Exit: Exploring the Equilibrium Correspondence Using the Homotopy Method," *Quantitative Marketing and Economics*, Vol. 10, 2012, pp. 197-229.

Christian, John, An *Overview of Past Proposals for Military Retirement Reform*, Santa Monica, Calif.: RAND Corporation, TR-376-OSD, 2006. As of March 27, 2013: http://www.rand.org/pubs/technical_reports/TR376.html

Defense Advisory Committee on Military Compensation, *The Military Compensation System: Completing the Transition to an All-Volunteer Force*, April 2006.

Department of Defense, *Report of the Tenth Quadrennial Review of Military Compensation*, Office of the Under Secretary of Defense (Personnel and Readiness), 2008.

Goldberg, Matthew, *A Survey of Enlisted Retention: Models and Findings*, Alexandria, Va.: Center for Naval Analyses, CRM D0004085.A2/Final, November 2001.

Gotz, Glenn, "Comment on 'The Dynamics of Job Separation: The Case of Federal Employees,'" *Journal of Applied Econometrics*, Vol. 5, No. 3, 1990, pp. 263-268.

Gotz, Glenn A., and John McCall, *A Dynamic Retention Model for Air Force Officers: Theory and Estimates*, Santa Monica, Calif.: RAND Corporation, R-3028-AF, 1984. As of March 27, 2013: http://www.rand.org/pubs/reports/R3028.html

Gustman, Alan L., and Thomas L. Steinmeier, "The Social Security Early Entitlement Age in a Structural Model of Retirement and Wealth," *Journal of Public Economics*, Vol. 89, No. 2-3, February 2005, pp. 441-463.

Hosek, James, Michael G. Mattock, C. Christine Fair, Jennifer Kavanagh, Jennifer Sharp, and Mark E. Totten, *Attracting the Best: How the Military Competes for Information Technology Personnel*, Santa Monica, Calif.: RAND Corporation, MG-108-OSD, 2004. As of March 27, 2013: http://www.rand.org/pubs/monographs/MG108.html

Hotz, V. Joseph, and Robert Miller, "Conditional Choice Probabilities and the Estimation of Dynamic Models," *Review of Economic Studies*, Vol. 60, No. 3, July 1993, pp. 497-529.

Keane, Michael P., and Kenneth I. Wolpin, "The Career Decisions of Young Men," *Journal of Political Economy*, Vol. 105, No. 3, June 1997, pp. 473-522.

Mattock, Michael G., and Jeremy Arkes, *The Dynamic Retention Model for Air Force Officers: New Estimates and Policy Simulations of the Aviator Continuation Pay Program*, Santa Monica, Calif.: RAND Corporation, TR-470-AF, 2007. As of March 27, 2013:
http://www.rand.org/pubs/technical_reports/TR470.html

Mattock, Michael G., James Hosek, and Beth J. Asch [MHA], *Reserve Participation and Cost Under a New Approach to Reserve Compensation*, Santa Monica, Calif.: RAND Corporation, MG-1153-OSD, 2012. As of March 27, 2013:
http://www.rand.org/pubs/monographs/MG1153.html

Rust, John, "Structural Estimation of Markov Decision Processes" in Robert Engle and Daniel McFadden, eds., *Handbook of Econometrics*, Vol. IV, Elsevier Science B.V., 1994, pp. 3082-3143.

Tilghman, Andrew, "Panetta Says He Will Seek to Protect Benefits," *Army Times*, August 19, 2011. As of September 7, 2012:
http://www.armytimes.com/news/2011/08/military-leon-panetta-interview-benefits-iraq-deployments-081911w/

Train, Kenneth E., *Discrete Choice Methods with Simulation,* Cambridge, UK: Cambridge University Press, 2003.

Van der Klaauw, Wilbert, "On the Use of Expectations Data in Estimating Structural Dynamic Choice Models," *Journal of Labor Economics*, Vol. 30, No. 3, July 2012, pp. 521-554.

Van der Klaauw, Wilbert, and Kenneth Wolpin, "Social Security and the Retirement and Savings Behavior of Low-Income Households," *Journal of Econometrics*, Vol. 145, No. 1-2, July 2008, pp. 21-42.

Warner, John, *Thinking About Military Retirement: An Analysis for the 10th QRMC*, Alexandria, Va.: Center for Naval Analyses, CRM D0017798.A1.Final, March 2008.